T0250422

The historical development of the social sciences ⸺ ⸺ frequent and fierce debates on the rules of scientific methodology. Even the most general criteria – which are generally agreed upon in the natural sciences – are emphatically disputed in the social sciences. In this book that phenomenon is more closely examined, using the historical evolution of twentieth-century Dutch psychology as a representative case of Western social science.

The existing variability of "methodological styles" in the development of the social sciences is illustrated by detailed historical material. Moreover, that material is used in the search for an answer to the main question of the book: how to understand the profound flexibility of standards in the social sciences. To that end, the historical scope of the study is restricted to the relatively small domain of psychology in the Netherlands, which offers a miniature image of international Western psychology. The author addresses not only "cut-and-dried" methodological criteria in manuals, but she carefully traces how certain rules in particular groups or periods could become the undisputed prescriptions for doing research.

Using published books and articles as well as interviews and archival material, Trudy Dehue analyzes the rejection, reception, and adjustment of German and American methodological styles in Dutch psychology, the rise and fall of handwriting analysis in assessment psychology, phenomenological psychology in the first postwar decades, and the "neurotic paradox of clinical psychology." She reports of psychologists calling colleagues "riffraff and cranks" or "semi-intellectuals who take lack of clarity for profundity" and reproaching each other for "undermining respect for men" or "discreditable dogmatism." The main thesis the author develops and tests is that in a given methodological style, apart from the *scientific* identity of the discipline, a *social identity* is at the same time given expression. Trudy Dehue argues that the degree to which a methodological style is accepted and maintained varies with the contemporary adequacy of the social identity expressed in it.

Changing the rules

Cambridge Studies in the History of Psychology

GENERAL EDITORS: WILLIAM R. WOODWARD AND MITCHELL G. ASH

This new series provides a publishing forum for outstanding scholarly work in the history of psychology. The creation of the series reflects a growing concentration in this area by historians and philosophers of science, intellectual and cultural historians, and psychologists interested in historical and theoretical issues.

The series is open both to manuscripts dealing with the history of psychological theory and research and to work focusing on the varied social, cultural, and institutional contexts and impacts of psychology. Writing about psychological thinking and research of any period will be considered. In addition to innovative treatments of traditional topics in the field, the editors particularly welcome work that breaks new ground by offering historical considerations of issues such as the linkages of academic and applied psychology with other fields, for example, psychiatry, anthropology, sociology, and psycholanalysis; international, intercultural, or gender-specific differences in psychological theory and research; or the history of psychological research practices. The series will include both single-authored monographs and occasionally coherently defined, rigorously edited essay collections.

Also in this series

Constructing the subject: Historical origins of psychological research
KURT DANZIGER

Metaphors in the history of psychology
edited by DAVID E. LEARY

Crowds, psychology, and politics, 1871–1899
JAAP VAN GINNEKEN

The professionalization of psychology in Nazi Germany
ULFRIED GEUTER

Changing the rules

Psychology in the Netherlands, 1900–1985

Trudy Dehue
University of Groningen

CAMBRIDGE
UNIVERSITY PRESS

CAMBRIDGE UNIVERSITY PRESS
Cambridge, New York, Melbourne, Madrid, Cape Town, Singapore,
São Paulo, Delhi, Dubai, Tokyo, Mexico City

Cambridge University Press
The Edinburgh Building, Cambridge CB2 8RU, UK

Published in the United States of America by Cambridge University Press, New York

www.cambridge.org
Information on this title: www.cambridge.org/9780521144872

© 1995 Trudy Dehue
© 1995 The revised English language translation Cambridge University Press

Originally published in Dutch as *De regels van het vak*
by Uitgeverij en boekhandel Van Gennep BV, Amsterdam, 1990
First published by Cambridge University Press 1995

This publication is in copyright. Subject to statutory exception
and to the provisions of relevant collective licensing agreements,
no reproduction of any part may take place without the written
permission of Cambridge University Press.

First paperback printing 2010

A catalogue record for this publication is available from the British Library

Library of Congress Cataloguing in Publication data
Dehue, Trudy, 1951–
[De Regels van het vak. English]
Changing the rules : Psychology in the Netherlands, 1900–1985 / Trudy Dehue.
p.   cm. – (Cambridge studies in the history of psychology)
Includes bibliographical references and index.
ISBN 0-521-47522-8
1. Psychology – Methodology – History.   2. Psychology – Netherlands –
History.   3. Social sciences – Methodology – History.   I. Title.
II. Series.
BF38.5.D3513   1995
150′.1 – dc20   94-2772
                                                                                CIP

ISBN 978-0-521-47522-8 Hardback
ISBN 978-0-521-14487-2 Paperback

Cambridge University Press has no responsibility for the persistence or
accuracy of URLs for external or third-party Internet Web sites referred to in
this publication, and does not guarantee that any content on such Web sites is,
or will remain, accurate or appropriate.

Contents

v

Preface

This book, like its subject, is also the product of a long history. In this history, many people and institutions, all of whom I would like to thank for their effort and support, have played a role. From 1985 to 1989 I carried out research in the section Foundations and History of Psychology at the University of Groningen. Pieter van Strien and Gerard de Vries were my much-valued supervisors in the writing of the Dutch-language dissertation that was the result of this research. Various people – who themselves helped to make the history of Dutch psychology – provided me with information at that time and various archivists also came up with unpublished material for me (the names of the people I interviewed and the archives I visited are given in the appendix). Uitgeverij Van Gennep in Amsterdam, who published my thesis in 1990 in a trade edition, ensured that the volume looked splendid and that it received a much wider distribution than is the normal lot of dissertations.

After that, numerous private reactions from colleagues and students, as well as the writers of fourteen reviews in Dutch-language journals set me thinking again about the subject. When the Netherlands Organization for Scientific Research was prepared to finance a translation into English – my special thanks go to Annemarie Bos for the friendliness with which she provided information and the efficiency with which she handled various matters – I decided to rewrite the text on a substantial number of points. But it was not only reactions from Dutch scholars that were helpful in the further development of my thinking. Kurt Danziger (who even reads Dutch), Mitchell Ash, and Bill Woodward also gave me very valuable suggestions. They provided references for more international comparison and they pointed out aspects of the original story that were so oriented to a typical Dutch public that they could not be followed by others, or would not be interesting enough to a non-Dutch audience.

vii

In this way Chapter 1 was largely rewritten and expanded, Chapters 2 and 3 have been made clearer, Chapter 4, in a rigorous attempt to sharpen the argument, became much shorter, Chapter 5 has been made clearer, and the Epilogue has also been largely changed. Of course, this process of rewriting took longer than I had anticipated. In the two and a half years during which I devoted the largest part of my research time to it, I was appointed to the Department of Science and Technology Dynamics at the University of Amsterdam, which offered a very stimulating and pleasant environment. I could with justice mention the names of all my Amsterdam colleagues here, but there are simply too many. For me they are best represented by Olga Amsterdamska, Stuart Blume, Ad Prins, and Paul Wouters. During this period I also worked in the section Foundations and History of Psychology at the University of Groningen, which under the direction of Pieter van Strien has for many years been my inspiring and familiar intellectual homefront. Here, too, students and colleagues offered a congenial intellectual and social atmosphere. From among them I would like to acknowledge in particular Peter van Drunen, head of the Archives of the History of Dutch Psychology, as a good colleague and friend.

Naturally, I must also mention the educational and enjoyable conferences of the Society for the History of the Social and Behavioural Sciences (Cheiron-Europe). This society offers a place for the international exchange of ideas that is indispensable for the writing of national histories too. Michael O'Loughlin translated the book without getting really angry when I changed a chapter again after he had carefully translated it. His many humorous reflections on language and culture added an entertaining extra dimension to our meetings. Alex Holzman at Cambridge University Press was always quick to answer any e-mailed question. And I want to express my great appreciation to Helen Wheeler who took care of the final editing. And I am very indebted to Anneke Pieterman for correcting proofs. So much support from others creates expectations. I can only hope that the readers of this new edition and above all those who have worked directly on behalf of the book will think that I have made good use of the help offered.

Abbreviations

Icip	Institute for Clinical and Industrial Psychology in Utrecht (*Instituut voor Clinische en Industriële Psychologie*)
NIP	Dutch Association of Psychologists (*Nederlands Instituut van Psychologen*). Known as NIPP before 1967.
NIPP	Dutch Association of Practicing Psychologists (*Nederlands Instituut van Praktizerende Psychologen*)
NRC	*Nieuwe Rotterdamse Courant*
NTG	*The Dutch Journal of Graphology (Nederlands Tijdschrift voor de Grafologie)*
NTP	*The Netherlands Journal of Psychology (Nederlands Tijdschrift voor Psychologie)*
NVB	Dutch People's Movement (*Nederlandse Volksbeweging*)
NVP	Dutch Association for Psychotherapy (*Nederlandse Vereniging voor Psychotherapie*)
PAI	Psychoanalytical Institute in Amsterdam
PvdA	Labor Party (*Partij van de Arbeid*)
TRC	Test Research Committee (*Test Research Commissie*)

Introduction

In the history of the social sciences the quest for the correct rules of science followed very diverse routes. The historical development of these sciences has been marked by frequent and fierce debates on their "methodology." Even the most general criteria, about which there seems to be agreement in the natural sciences, are here emphatically disputed.

In this book that phenomenon is subject to closer study. The existing variability of "methodological styles" in the development of the social sciences is illustrated by detailed historical material. Moreover, that historical material is used in the search for an answer to the main question of the book: How should such a profound flexibility of the standards in the social sciences be understood?

Not just the "cut-and-dried" methodological criteria in manuals are studied, but the ways in which certain rules in particular groups or periods could *become* the undisputed prescriptions are carefully traced. To that end, the historical scope of this study is deliberately restricted to the relatively small domain of psychology in the Netherlands. Psychology has always been strongly methodology oriented, and the Netherlands is not only conveniently small, but because it is wedged geographically between many great powers, its inhabitants traditionally are very internationally oriented. Dutch psychology therefore offers a miniature image of international Western psychology: It is small enough to be studied at a detailed level and nevertheless varied enough to be representative of the discipline's variability.

Chapter 1 gives a brief historical overview of methodology in Dutch psychology that describes phases of an orientation to the German *Geisteswissenschaften,* a Continental-European phenomenological approach and an Anglo-Saxon empirical-analytical style. This chapter explains why this study does not focus on what researchers *actually do* but what

1

prominent representatives of the field write down that they *should do*.
Also, the use of the concept of methodology is defended to encompass
every position taken by academically trained and prominent representatives with regard to requirements for qualitatively good psychological
work.

Next I discuss the option that methodological changes in the history of
the social sciences can simply be ascribed to a steady growth of insight
into the proper rules of science. If this is so, psychology in the Netherlands, which is now mostly oriented to a hypothesis-testing empirical-
analytical approach, slowly but surely seems to have come to realize how
the job must be done if it is to be carried out in a scientifically correct
manner. However, I argue that historical work, which shows such a
process of growth, generally does so by declaring, implicitly or explicitly,
the contemporary state of affairs as its criterion. In the presentist or
finalistic accounts that arise in this way, an objective linear growth of
insight from the past to the present is suggested, whereas in fact the line
is actually drawn by the historian from the present to the past.

This leads to the question of whether the image of cognitive progression can be maintained in a historical study where the contemporary state
of affairs does *not* function as a criterion. In principle, in a nonfinalistic
approach, too, constellations of now-antiquated rules might turn out to
have been inadequate and a new methodological style could turn out to
have been more suitable. However, such a conclusion would imply that
there were periods in the history of Dutch psychology when academically
trained and experienced practitioners of the field, who furthermore were
aware of the better methods via the professional literature at home and
abroad, persisted for decades in making errors that were against their
own interests. Because such an unlikely picture is convincing only when
it has turned out to be inevitable, I decided first to develop an alternative
option and to check its plausibility against the most important episodes
and groups in the history of Dutch psychology.

On the basis of an analysis of four characteristics of social scientific
methodology, I develop an alternative hypothesis in the last sections of
Chapter 1, namely, that in the course of its history Dutch psychology
affiliated itself with different social groups having different ideas on social
relationships and that this sociocultural pluralism found its expression in
its variable methodology. Thus, in other words, it is assumed that in a
methodological style, apart from the *scientific* identity of the discipline, a
social identity is at the same time also given expression and that studying
the process in which a discipline or a group disposes of an old scientific
identity and assumes a new one demands the discovery of the changing
social identity of that discipline or group. The first chapter concludes

with a discussion of the differences between these presuppositions and those of "contextualist" historiography, which searches for the external (and internal) "influences on" science. I present reasons why the question posed in this book is not what factors influence methodology, but what urges a scientific community to create or subscribe to a specific scientific identity.

Chapter 2 concerns the methodology of psychological assessment for personnel selection and vocational guidance. Emerging under the name of "psychotechnics" in the 1920s, it remained the dominant branch of Dutch psychology until the 1950s, and to this day it is an important field of psychological activity. The crucial example in this chapter is the debate about characterological interpretation of handwriting ("graphology") as a psychodiagnostic instrument. This debate, which blazed fiercely in the fifties and sixties, confronted two distinct methodological styles. The first style, known as *geisteswissenschaftlich* – which included graphology – lost the battle. From the mid-sixties onward, the "empirical-analytical" style became almost commonplace.

To answer the question of how the rules came to change in this case, first the issue is raised of why the former empathizing methods, which now sound so unconvincing, were once able to seduce people and institutions into spending money and effort on psychological diagnostics. Following from that question, the second asks: What led to the empathizing methods *loss* of persuasiveness and to the acceptance of quantifying empirical-analytical methodology?

I demonstrate that the methodological rules of the former psychotechnicians were tailored to an interpretation of the question of professional eligibility as a request for guidance in finding life fulfillment. This interpretation was developed in the welfare institutions where psychotechnicians were employed, and it was viable as long as the so-called pacification democracy, with its image of harmonious labor relations, persisted in the Netherlands.

With the politicization and polarization of Dutch society in the fifties and sixties, the role of the assessment psychologist as a social worker could not be maintained. In place of the model of the social worker, the psychologists presented that of the businesslike, detached middleman in manpower. The profession became governed by a new set of methodological rules that proved to be appropriate for this new role.

So this history contradicts the view that the former psychotechnicians did not yet properly understand how their profession should be "scientifically" practiced, whereas the postwar psychologists gradually came to understand this better. I conclude that the changes in the methodology of assessment psychology were not the consequence of an accumulation of insight into "the" independent criteria for the correct approach, but rather were the result of the fact that the profession itself changed as a component of a changing society. The change of the *scientific identity* of the discipline pointed to a change of its *social identity*.

Chapter 3 deals with phenomenological psychology at the University of Utrecht in the period of 1945–70, which became known as "the Utrecht School" in psychology and attained a strong position in the international "phenomenological movement." First, the methodological aspects of the Utrecht phenomenological psychology are worked out. Next, I demonstrate that this methodology can be understood properly only if one realizes that it was not Husserlian phenomenology but Max Scheler's personalistic philosophy that was the main intellectual basis of its viewpoints. That point raised the question why the members of the school named their approach phenomenological psychology and not personalistic psychology. The fact that *personalistic socialism* appears to have been the source of the school's social ideals answers that question and elucidates how it was possible that this psychology could flourish precisely in the first postwar period in the Netherlands, whereas in the 1960s it was quickly forced back by the statistical empirical-analytical approach.

Personalistic socialism thrived during the Second World War in the Netherlands and found a strong following in some circles after the war, but it soon became controversial and almost completely lost its cogency in the sixties. Thus, the thesis is illustrated again that subscribing to particular professional rules involves choosing specific positions on the level of social relations. I show, this time on the basis of the Utrecht phenomenological psychology, that the degree to which a methodological style is accepted and maintained varies with the contemporary adequacy of the social identity expressed in it.

In Chapter 4 the example of clinical psychology from the 1920s to the present provides a more complicated case. Along with a gradual closure of the debates on what is to be called science, there is a *remaining* uneasiness. This uneasiness was expressed by Johan T. Barendregt, the main character of the chapter, in the metaphor of "the neurotic paradox."

Barendregt entered the field of clinical psychology in the 1950s and in the 60s became a professor at the University of Amsterdam. The central dilemma throughout his career was that if he regarded a research project as methodologically correct, he believed it was not relevant for life outside of science and hence for clinical practice, but if he considered a project to be relevant, he criticized it for not being methodologically correct. To him this dilemma, more often expressed in the international social sciences, became so serious that living up to methodological prescriptions became a neurotic paradox: "a romantic image which ruins your life."

As Barendregt with his paradox described a problem of Dutch clinical psychology as a whole, he is presented as the personification of the dilemma of the subdiscipline. The central question of the chapter is how clinical psychology in the Netherlands fell into this extreme dilemma of rigor versus relevance and why its doubts did not go away.

I argue that the dilemma is not the inevitable outcome of an essential and universal incompatibility, but the consequence of an *incompatibility of social identities* that emerged historically in clinical psychology's laborious process of emancipation with regard to psychiatry. In the course of time, clinical psychologists fought for their independence from the psychiatrists who employed them, not only by introducing their own methodology borrowed from industrial assessment psychology, but also by assuming their own social identity in the context of mental health care. And *that* social identity was very different from the identity traditionally bound up with assessment psychology. I conclude that the neurotic paradox of clinical psychology in the Netherlands is another example of a methodological issue that cannot be properly understood without taking into account the social identity that is expressed in a methodological style.

The various themes of the book come together in Chapter 5, and a next step is taken. I raise the objection that showing the rationality of several methodological styles in their own social setting does not necessarily mean that all of these approaches can also be called *scientific*. It is still possible perhaps to speak of Dutch psychology as gradually becoming scientific in the sense that it has finally linked up with the image of science that just happens to be the *internationally current* one. Looked at in this way the history described in the previous chapters did not in fact refute the image of a steady growth of insight into the proper rules of science, but showed only that certain social circumstances provide the necessary conditions for the real scientific approach to flourish.

However, first I argue that this position simply denies the many methodological styles in the international social sciences that even today distance themselves from empirical-analytical thinking. Empirical-analytical rules are not as internationally current as some may wish them to be. Moreover, this position overlooks the fact that even *within* the empirical-analytical realm, the unanimity it assumes is not to be found.

In this chapter, the stance of international agreement on empirical-analytical rules turns out to be untenable too. On closer philosophical consideration the mainstream of Dutch psychology appears to adhere to a particular individual variation within the international empirical-analytical framework. Specifically, the rule that the Dutch themselves consider to be the *key* notion of scientific methodology is not to be found in either the closest philosophies of science (logical empiricism and falsificationism), or international handbooks of social scientific methodology.

Furthermore, this Dutch variation appears to be related directly to the changing social identity of psychology in the Netherlands. The methodological rule of Dutch psychology that is most taken for granted, as shown at the end of this chapter, is the product of the problems, unique to a certain extent, with which Dutch psychologists were confronted in the course of history and with which they would not have been confronted if their discipline had not had its own changing social role to play in its own changing society.

In this way, finally, the standard view of a growing insight into an abstract general rationality is exchanged for a different picture. That is the picture of social scientists who, though using methodology that is available elsewhere, nevertheless participate actively in their own society and in accordance with this define the rules of their discipline. I also conclude that the former deviations from today's standards should not be ascribed to intellectual confusion in the past. The alternative image is that psychology in the Netherlands affiliated itself to many different social groups and that a sociocultural pluralism is expressed in its variable methodology.

The epilogue discusses the corollary that even if researchers in everyday practice always behaved according to the rules, that would not guarantee that their work would be culturally independent. This, of course, is by no means the last word on methodological matters. On the contrary, in the end, the symmetrical approach in this study was only a means on behalf of the contemporary methodological debate itself. In the epilogue, some consequences for this debate are briefly indicated.

1

The variability of methodological standards
in the social sciences

In the empirical science studies of recent decades, a completely different picture has been drawn of the development of scientific knowledge than that portrayed by philosophers of science and the majority of scientists themselves. Both groups ascribe scientific progress to following the correct *methodological rules*. However, on the basis of empirical investigations, researchers in the field of science studies claim that this picture has little descriptive or empirical adequacy. They argue that agreement with regard to methodological criteria in fact exists at best on an extremely general level and that in practice these general criteria are interpreted in diverse ways, or even that researchers do not at all keep to methodological rules in carrying out their work. The development of scientific knowledge, many argue, is actually a social process.[1]

If the contribution of methodological rules to the development of scientific knowledge is negligible, as some have reasoned, it must be explained which objectives drawing up methodological criteria actually does serve. Various *rhetorical* functions of methodology are revealed: Methodological arguments are said to reinforce retrospectively one's rightness, to impress the outside world, or to legitimize preconceived and culturally determined differences.[2]

Apart from this, through the years studies have been carried out on the historical development of the rules of science.[3] Methodological rules do not determine the most important theoretical and empirical developments, but were actually determined by them, it was argued.[4] Others

Parts of this chapter have been published in English in T. Dehue, "Why Methodology Changes: Transforming Psychology in The Netherlands I," *History of the Human Sciences* 4 (1991):335–51; T. Dehue, "Why Does Methodology Change Over Time? A Theoretical View and its Implications," in H. J. Stam, et al, eds., *Recent Trends in Theoretical Psychology, Vol. III* (New York: Springer Verlag, 1993), pp. 495–502.

advanced a sociological approach to the historical multiplicity of criteria, which consisted not of scientific theories, but the culturally steered consensus of researchers or their image of the ideal mutual relations that determine which rules have the most value ascribed to them.[5]

Once methodology had been reduced to the "fifth wheel on the wagon," however, not *much* interest remained in the characteristics and development of the rules. The number of empirical studies into the functions of methodology, or into the question of how a particular conglomerate of rules *is constructed* or *established,* lags far behind the number of investigations that only focus on falsifying the functions methodology is said to fulfill by philosophers and scientists themselves. In addition, the existing studies are almost exclusively oriented to diverse quantitative research styles. Little attention is paid to the social and behavioral sciences (henceforth for the sake of brevity to be referred to as "the social sciences").[6] And when methodology in these sciences is empirically and historically investigated, the focus is mostly on those methods and techniques that are derived from the natural sciences.[7]

However, in disciplines like sociology, psychology, and political science, the quest for the correct working method led to even more diverse routes than was the case in the natural sciences. Also widely discussed in these sciences are general criteria such as the preference for quantitative statements over qualitative or predictive accuracy, about which, according to some, there is agreement in the natural sciences.[8] It has by no means always been, nor is, a matter of course for all academic practitioners of the social sciences to believe that one is allowed to speak of "science" only when general laws are sought, predictions are made, and quantitative values are ascribed to observable phenomena. Such conditions were, and continue to be, emphatically disputed by phenomenologists, hermeneuticists, ethnomethodologists, constructivists, and the like. They reject the "empirical-analytical" way of thinking, as I will call the broad category under which various quantifying approaches fall.

In this book the phenomenon in the social sciences that there are differences in thinking among groups or between time periods on even the level of the most general rules will be subject to closer study – not only because it is a relatively unexplored topic, but also for reasons to be extensively elucidated later in this chapter. The existing variability of methodological "styles" in the development of the social sciences is illustrated on the basis of detailed historical material. Moreover, that historical material is used in the search for an answer to the main question of the book: that is, how such a profound flexibility of the standards in the social sciences should be understood, or more specifically, how a particular scientific community comes to create or endorse a particular methodological style.

Psychology in the Netherlands as a case study

In such a project, it is not enough to study the cut-and-dried methodological criteria in manuals. As Latour put it, it is necessary to work back from the law book to the parliament when the laws were still bills.[9] It must be discovered how in a particular group or period certain rules could *become* natural conditions. This requires a level of detail that is achievable only by restricting one's research to a small domain.

Psychology in the Netherlands is a field particularly suitable for my purpose. From its beginning as a university discipline, psychology has been strongly methodologically oriented. Moreover, Dutch psychology is particularly suitable if only because the Netherlands is so conveniently small (some 34,000 square kilometers – 13,000 square miles – of land, with at the moment approximately 14.9 million inhabitants,[10] including around 20,000 psychologists).[11] The Dutch, moreover, wedged in between German-, English-, and French-speaking great powers, are traditionally very internationally oriented. They have always been aware of what is written in other countries, and academics in particular did not have to wait for Dutch translations of foreign-language materials. As the Dutch labor historian A. J. C. Rüter remarked about his countrymen:

> We sit and we have always sat at a crossroads in Europe, and our borders have always been open to traffic – intellectual traffic, too. Our civilisation has therefore continually absorbed elements which it borrowed from the great cultural and social currents traversing Europe. If the labour movement borrowed many features from abroad, this does not make it un-Dutch. On the contrary, I view this ability to absorb as a typical Dutch characteristic, at least to the extent that the labour movement refashioned the borrowed elements into an entity which, as an entity, carried a specific Dutch imprint.[12]

This volume will show that these characteristics hold not only for the labor movement but for the methodology of Dutch psychology as well.[13] From the profession's beginning as a university discipline, methodological standards from Germany, England, the United States, and to a lesser extent also from the less linguistically close France, easily got through to the psychologists in the Netherlands. They assumed those standards directly, or placed them beside their own convictions and adjusted them according to their requirements. Thus, in a small geographical area, over a period of hardly a century, many different methodological styles were in competition with each other. In contrast, in a study of psychology in, say, the United States or England, the historian mostly encounters methods oriented to the natural sciences,[14] whereas historically Dutch psychology, as well as French and German psychology, were also open to Continental-European philosophy. Therefore the Netherlands offers a miniature image of international Western psychology, small enough to be studied at a detailed level and nevertheless varied enough to be representative of the discipline's variability.

In his publications on the "phrenology disputes" in early nineteenth-century Edinburgh, the historian of science Stephen Shapin concentrated on just one town.[15] These studies graphically illustrate that restriction to a small geographical area also has some extra advantages. As the studies show, spatial nearness more easily leads to controversy and controversies in particular are very instructive for science studies. Because traits are never as clearly revealed as they are in mutual clashes, they offer an opportunity to obtain clear pictures of all the parties involved. Moreover, the "reopening" of what are now closed debates makes us realize that what nowadays seems obvious was once by no means so evident and clear.[16] For example, the sociologist of science Harry Collins has compared currently known self-evident truths in the sciences to ships in bottles. Just as a ship was not always there in its completed form, but was once nothing more than a pile of sticks, scraps of cloth, and dabs of glue *outside* the bottle, scientific facts and insights are not givens that have been discovered but the final result of a series of operations, decisions, and events. The description of periods of instability and uncertainty in the history of a scientific field leads us, to begin with, to *questioning* how the ship got in there. With an analysis of the process whereby the debate reached its present outcome, it can then be shown how the sticks, the rags, and the lengths of rope ended up as a ship in the bottle.

Particularly the fact that spatial nearness leads to controversies is why psychology in the Netherlands is a very suitable case for my aims. Indeed, many strong confrontations will be reported in the coming chapters. Meanwhile, I will not attempt to give a complete historical overview of everything that has been presented in Dutch psychology on the level of methodology. Rather, the most important episodes and themes from its history will be used to test the premises on which methodological changes and consensus formation take place, a theme to be developed in the last section of this chapter. Let me first give a brief and broad overview of this history of methodology and add a more precise definition of the concept of "methodology."[17]

A historical outline and the concept of methodology

In 1933 Charles Spearman, a statistician and professor of psychology in London, sent a "call" to the various European psychology journals. On the basis of correlations between intelligence tests and of factor analysis, Spearman had previously developed the idea that all forms of intelligence can be reduced to one single factor "g" ("general intelligence").[18] Now he asked the psychologists of Europe to draw up lists of what they considered the traits that most determined personality. They were supposed to send the lists to the University of Chicago. There, the statistical correlations and differences between all traits mentioned would be studied and, as the call expressed it, the correctness of the resulting factors would then be empirically assessed. This plan would result in a definitive solution to the question of whether and how human personality can be defined by a limited number of features and traits. Besides, the project was supposed to rescue European psychology from the crisis in which, in Spearman's view, it found itself. This crisis was formed by the existing multiplicity of mutually contradictory theories, which according to Spearman was in its turn a consequence of the neglect of empirical verification.

If such an appeal had reached the Netherlands some years earlier, it would possibly have met with a response there. In the first decades of this century, the Groningen philosopher and psychologist Gerard Heymans studied the individual differences between people. Heymans had also looked for general factors defining personality. He collected empirical material with the help of questionnaires and biographies. In his "temperamental typology" he defined eight personality types on the basis of varying combinations of the positive and negative poles of three main dimensions.[19]

However, in the course of the 1920s the tide had turned. Heymans's former students considered his "elementaristic" methods no longer adequate. They were more inspired by German thinkers from the holistic *geisteswissenschaftliche* tradition.[20] Instead of tracing and testing general laws in individual diversity, they advocated the intuitive approach to every "unique person." The psychiatrist and psychologist Lammert van der Horst, who was a professor at the Calvinist "Free University" in Amsterdam, had also noted a crisis in psychology. In a 1933 article in the *Netherlands Journal of Psychology* (*Nederlands Tijdschrift voor Psy-*

chologie, hereafter *NTP*), he based the same diagnosis on opposing phenomena: In psychology too much emphasis was being placed on "the measurable and quantifiable" and on an "atomistic point of view." As a remedy, Van der Horst emphasized "the insoluble basic relation": "I know something, I experience something."[21]

In the same year, when Spearman's appeal arrived at the *NTP,* the board grasped the opportunity once again to distance itself strongly from the sort of psychology that Spearman advocated. It did publish the call,[22] but in a critical postscript it argued that the existence of theories of personality did not indicate a crisis, but was actually a sign that "in any case man is himself as a person and nothing else." Statistical psychology had "no validity with regard to real psychological matters." Only the "genius and vision of an individual" could reveal "the finer relief of human personality." Readers were advised to regard this "Anglo-American concern about our psychology" as a sublimation of a concern that actually relates to American psychology itself. "That is probably the reason for its naivety," surmised the editors, "because do people on the other side of the ocean really think that scholars like Spranger, Jung and Klages, these theoreticians of personality par excellence, would sit on a standardization committee for the traits of human personality? That would be like Rembrandt, Michelangelo and Frans Hals together trying to define the 'true' painting."[23]

Since then, the differences of opinion on the conditions, norms, and rules of the practice of psychology have not ended in the Netherlands. From the University of Amsterdam, a crisis was declared again in 1950 by Professor Adriaan D. de Groot, and a renewed empirical-analytical approach was then put forth.[24] The central norm in this would have to be the prediction of research results. According to De Groot, interpretative methods could be used only to conceive *hypotheses* about an individual. Next, objectively testable statements should be derived from these hypotheses. In this way De Groot combined elements from both the intuitive German-oriented *geisteswissenschaftliche* style and the American-oriented statistical style. The role of intuition was preserved in the creation of hypotheses about an individual and this individual was not necessarily compared to a group norm, but could still be studied "in his uniqueness." Nevertheless, the criteria for sound psychodiagnostics were derived mainly from the statistical tradition. What counted first was the calculation of the degree of correlation between a quantified judgment and a quantified criterion (in other words, the calculation of the "validity" as it was then called in imitation of Anglo-Saxon test psychology). Furthermore, the degree to which a method yields stable results (that is, its reliability) was now expressed in a number and accepted as a crucial aspect of assessment psychology. Only in this way would psychologists arrive at scientific empirical statements, De Groot and his colleagues argued.

In the 1950s and 60s many debates and conflicts centered on the question of how psychological research should be done. As I shall describe in this book, in meetings, scientific journals, and in the national press, the parties hurled at one another reproaches of "unscientific behavior," "violation of humanistic principles," "terminological vagueness," "prejudice," or even the notion of "treating people like kilos of iron."

In the same period, a "phenomenological psychology" was proposed by psychologists at the University of Utrecht. These psychologists rejected both the *geisteswissenschaftliche* and the empirical-analytical methods with the argument that both made people into objects. In place of these methods, phenomenology suggested the "desire-free encounter" with the fellow man, as the only real way to get to know "the other." In order to become a phenomenological psychologist it was not necessary to master methods and techniques, but rather it was the psychologist's task to acquire the proper attitude about life through education. This Utrecht psychology had a large following not only among Dutch psychologists, but also became authoritative in the international phenomenological community (see Chapter 3).

In the course of the 1960s, however, the predictive empirical-analytical approach quickly gathered support while that of phenomenology dwindled. In 1961 De Groot published a book entitled *Methodologie,* in which he expanded his ideas on psychodiagnostics into a methodology for the behavioral sciences in general. This book achieved its sixth impression in 1970, with the publication of the English edition taking place in 1969.

The school of thought, expressed in Van der Horst's aphorism, "If I know something, I experience something," was now almost generally exchanged for De Groot's exhortation: "If I know something, I can predict something; if I cannot predict anything then I know nothing."[25] According to this view the psychologist had to train himself extensively in the methods and the statistical techniques of hypothesis-testing research. Apart from De Groot's colleagues at the University of Amsterdam, psychologists at other universities in the 1960s also developed empirical-analytical standards and accompanying techniques. Some of them focused primarily on the methods of research into the individual differences between people. Others mainly defended nomological research in which the general laws of human actions and capacities were sought with the help of experimental methods.

In the 1970s, a younger generation espoused Maslow's "humanistic psychology" or a Marxist "critical psychology" with methods like "action research," which explicitly took socialist stands on human relations.[26] However, most new variants were developed within the empirical-analytical style. In Groningen, for example, the personality psychologist and methodologist Willem K. B. Hofstee, on the basis of a fallibility epistemology combined with the desire to remove "speculation and humbug" from the social sciences, developed a "betting model" for carrying out research in the social sciences.[27] At some universities, a movement arose in the ranks of the nomological line that oriented itself to systems theory, and at Nijmegen a "mathematical psychology" was developed as a more exact alternative to the common hypothesis-testing approach, because working with operationalizations of concepts was considered to be still far too interpretative.[28]

From the 1970s onward, this hypothesis-testing and empirical-analytical thinking therefore forms the natural context in which the leading psychologists in the Netherlands make their assessments and statements on scientific research. The few latter-day proponents of the "hermeneutic" or "constructivist" approaches, based on the *geisteswissenschaftliche* tradition, must defend the disputed right to existence of working methods that fall outside the borders of the empirical-analytical framework. Thus, in the Netherlands (as in the United States and many other countries) the concept of science among contemporary psychologists has now become connected with carrying out experiments and testing hypotheses to such an extent that the concept of methodology has become synonymous with the methods and techniques of empirical-analytical research.[29] However, the *historian* who follows such habits in the field he or she investigates is confronted with specific problems. If, in a study like the present one, the term "methodology" was reserved only for empirical-analytical standards, then only criteria for a scientific psychology that sufficiently resemble the contemporary ones could be included. The concept of methodology would not apply to views like those of the *geisteswissenschaftliche* or phenomenological psychologists discussed earlier, and in this way they would fall outside the scope of this project. It would, then, be incorrectly suggested that these latter perspectives have played no role in the creation of the present-day rules in Dutch psychology and it would be incorrectly suggested that contemporary methodology has historically unfolded as a single ongoing empirical-analytical line.

In this book therefore the concept of methodology is used not only for the methods and statistical techniques of empirical-analytical thinking, or for this style of thought in a broader sense, but also in its universality with regard to all answers given, namely in the past by academics to the

Figure 1. Map of the Netherlands with the university cities mentioned in this volume (M. Balyon, Graphical Design, Leiden).

question of how psychological research should be done. These answers are characterized by great diversity, in the sense that the *themes* that were considered relevant by the various groups or in various historical periods in relation to ways of research can diverge sharply. In discussing the postwar Utrecht phenomenology, for example, positions regarding the essence of human existence and concerning the *moral* conditions for real knowledge will appear to belong to methodology. Such ethical or philosophical-anthropological questions are not dealt with in empirical-analytical methodology. The latter mainly considers epistemological topics and its proponents concentrate on the development of methods for *eliminating* moral and "metaphysical" beliefs.

So, what one researcher considers to be one of the crucial elements of the rules of the discipline is considered by another not to belong there at all. For this reason it is not possible to ascertain precisely what the *essence* of methodology is or what the specific characteristics are of all methodological styles. Methodological ideas show mutual resemblances just as families do: Whereas each family member has a very individual nature and appearance with which he or she is distinguished from all the other members, at the same time character and appearance consist of a compilation of features, some of which the others also show. Just as it is not possible to point out a single feature that applies to all, and is characteristic of a family, it is not possible to define precisely the concept of methodology.[30] In this book therefore a somewhat vague definition of methodology is used deliberately. Methodology is *every position taken by academically trained and prominent representatives with regard to requirements for qualitatively good psychological work.* It discusses norms and rules for correct professional ways, which also indicate why, or under which conditions, the products of the discipline can lay claim to the designation "scientific." The term "methodological *style*" refers to a specific constellation of mutually related ideas on this.

The present as norm, and the present as starting point in writing the history of scientific disciplines

An explanation for the developments of psychological methodology in the Netherlands as just described, which at first sight seems plausible, is that the discipline simply went through a slow process of "scientification" there. Some side roads have been followed, but most psychologists in the Netherlands have by now understood how they must do their work if they want to do it scientifically. After a period of wandering, the great majority of Dutch psychologists are onto the right track.

Histories have been written, not only of psychology in the Netherlands, but also of other sciences, in which such an image unmistakably seems to come to the fore. The historical introductions to textbooks and manuals are the clearest example of this. Here, the steady maturation of the field is clearly visible. However, by now many objections have been raised against what is called the "whiggish" or "presentist" way in which such an image of linear progression is generally constructed.[31] These objections have even been so exhaustively described that many contemporary historians of science consider a single dismissive reference sufficient.[32] Anyone repeating the arguments against "whig history" nowadays would appear to be beating a dead horse.[33]

In the context of a study into the variability of methodological styles, however, clarity about this is of special importance. If the image of a cumulative growth does not appeal to many scholars in the field of science studies anymore, particularly when methodology is at stake, then it certainly does appeal to most scientific researchers themselves. If only because I hope at least some of them will be among the readers of this book, the arguments for and against the image of linear progression must be dealt with as carefully as possible. In addition, among historians and sociologists of science, under the superficial consensus there often appear to be profound differences of opinion about the precise meaning and consequences of a particular stand on the subject. Therefore, too, I will elucidate in what sense I am using the concepts of whig history and presentism, and I will then discuss some views often related to the antipresentist stand to which I do *not* subscribe.

As historians have often clarified, the term "whig history" was introduced in 1931 by the English historian Herbert Butterfield with the book *The Whig Interpretation of History*. Butterfield was referring to the political history in England, in which the dominant ideas of the Whigs were implicitly used as the norm in the study of the past.[34] He and others expanded the meaning of the term to include every form of history that uses the assumptions of the period in which the historians themselves live, or of the party they belong to, as a criterion for the selection and evaluation of earlier events.

Many scholars in science studies have shown that in the history of scientific disciplines in a similar way the current state of affairs often serves implicitly as the standard by which to arrange the past. The present-day image of what psychology (or sociology, or chemistry) "is," for example, serves as a criterion to determine who in the past could rightly claim to be working in the field and who can therefore be included in history as a colleague. Nor are the changing meanings of professional terms such as experiment,[35] objective, subjective, prediction, hypothesis or – as I have already mentioned – methodology, taken into account, but rather one's own definition is imposed implicitly or explicitly as a norm on the past. In this way the historian is simultaneously *reporter of* and *participant in* the historical struggle: The story of the debates is written from a viewpoint inside the debates.

The genre may not only be dismissed as "presentism" but also as "finalism." By constantly jumping ahead to a later state of affairs in developments when describing the past, the idea of a single valid viewpoint is conveyed, which in the course of history is slowly revealed by steadily removing all obstacles and mistaken ways of thinking. In the course of the narrative the image of a line of development is sketched that is not drawn by the author from the present to the past, but that indicates an objective path from the fresh first discovery of the correct scientific thought to the present.

It is not without reason that the historical introductions to manuals and textbooks provide prototypes of this sort of history. Whig histories support the identity of the field and they underpin the contemporary ideas with the foundations of a long tradition. In this fashion they can serve as a means to socialize students in the contemporary insights of the discipline. But anyone who wonders how it was possible that in a particular period certain methodological ideas were adhered to will get little from a presentist history. If the contemporary styles were not followed, then the ideas involved are simply regarded as wrong trails (or at best as expressions of "prescientific thought"). This kind of history does not ask what assumptions the parties involved based their reasoning on when taking up positions that now appear strange, nor does it address the kinds of questions and problems to which the parties were responding. If predecessors did work in the contemporary style, then there is even less reason to investigate how that came about: The ones who did so were merely the pioneers in the field. In this fashion, the fact that in the past methodological positions different from present-day ones were convincing is presented as the slow discovery and growth of rationality. Such a response to the question of how methodological styles change, however, is begging the question, because the question is *How it did come about* that the criteria of rationality do change.[36]

For some time there has been a popular alternative to this explanation of historical change as a process of linear or "internal" growth. "Externalistic" or "contextualistic" historiography does not regard the development of a scientific field as an autonomous process, but as steered, or even "also steered," by "external" ("social," "contextual") factors. On the basis of this idea empirical research has been conducted regarding the diverse internal and external "influences on" the development of science. In such studies the relative weight and mutual interrelatedness of the various internal and external factors usually remain unclear. Moreover, no less than presentist histories, this kind of research assumes an independent and static criterion for what science "is" and what it is not. This is because it needs a standard for the *division* into internal and external factors. As in presentist histories, contemporary boundaries are implicitly taken as invariable standards.

However, in the historical development of disciplines various definitions of science have been defended and fields have been variously demarcated. The "scientific identity" of phenomenological psychologists, for instance, differed considerably from that of empirical-analytical psychologists. What belonged to essential parts of science, regarded in terms of the first frame of mind, fell into extrascientific categories like religion and novel writing, when seen from the second. Anyone who now describes the influence of novel writing and religion as external factors elevates empirical-analytical thought to a standard (and whoever describes them as internal factors does the same with phenomenology). Divisions in internal factors and external factors cannot exist independently of the various methodological styles. Rather, labeling a factor as "internal" or "external" always is the *product* of a particular methodological style. The varying rules themselves constitute the varying differences between science and context.[37]

It is this process of constituting a new scientific identity by means of a new methodology that has to be studied to gain insight into the phenomenon of diverging methodological styles. In my view, the question to be posed is not what (external and internal) factors influence methodology, but what urges a scientific community to subscribe to or create a specific scientific identity in a particular period.

Nor do I subscribe to another alternative to whig history. "Presentist" criticism is connected by so-called historicists to the idea that historians should approach the past with complete detachment. They should eliminate all knowledge of what was to come from their memories and keep all hypotheses about how it has been from their minds. The task of the historian would then merely be to apply her- or himself to empathizing with a period to be studied. However, in this way the criticism of presentism threatens to turn into a plea for naive empiricism. Historians cannot completely detach themselves from their own time. Furthermore, historical studies are not possible without questions and ideas preceding them. As in research in general, questions and presuppositions give direction in the selection from the sea of potential data. Those who do not have surmises about the past will not know what to study. Attempting to meet the demands of total detachment will also quickly deprive the writing of history of its relevance to the present. Historical research will then result in a collection of antiquities without line or cohesion.

The idea that with antipresentism a naive-empirical historicist standpoint would be *inevitable* has led others actually to argue in favor of "presentist" history again. By this they mean history that proceeds on the basis of questions and assumptions derived from the present. Confusingly enough, they therefore argue for "presentism" by making a plea for positions comparable to those of (not naive-empiristic) "antipresentists."[38]

I myself use the concept of presentism in the pejorative sense for taking contemporary certainties of a discipline implicitly or explicitly as a *norm* for the past. The intention to avoid this as much as possible does not yet imply any ban on choosing contemporary ideas as the *point of departure* of a study. In the studies into psychological methodology to be presented here, I may also depart from the present. The difference is that no straight line is drawn from present views as a criterion to precisely those earlier standpoints that "already" resembled the criterion; rather, the paths are traced that resulted in the present-day point. I do not rule out in advance that straight lines will result in the latter manner also, but nor do I exclude in advance the idea that a route will become visible, one that was followed by many various actors on the basis of very different motives and with divergent goals. Then, this route will have run in many directions. From this perspective, the point where methodology stands today was not the goal from the beginning, but is the current *outcome* of the trajectory.[39]

Finally, the intention to take no part oneself in debates during the historical description of these debates does not imply an eternal ban on developing normative judgments. The outsider's position that comes with the rejection of presentist partiality does not prescribe universal aloofness. Although I do not write my narrative using one of the styles to be treated as a criterion or in other ways directly criticizing methodological stands, at the end of the book I briefly discuss some consequences of my analysis for the normative methodological debate.

In the next section I defend the importance of research in methodology. In doing so I also discuss four characteristics of the disciplinary rules. Arguing on the basis of these four characteristics, I finally present in the last section my conjectures about the process in which a discipline or a group within it assumes a new scientific identity.

The importance of research into social-scientific methodology

Why study the variability of methodological rules? As I mentioned in the first section, in the practice of scientific research general rules may be interpreted in various ways. Nor will the actual research process by any means always proceed according to imposed or self-imposed standards. As with people in general, there is unmistakably a difference between what scientific practitioners do and what they say they should do, or even what they claim to have done. The historian of science Jennifer Platt, who wrote revealingly on the discrepancy between norm and deed in sociology, therefore warned her fellow historians against "the easiness of reading statements of principle rather than observing the complexities and ambiguities of practice."[40] And Danziger in his scholarly book on the social aspects of investigative practice argues:

> Methodological prescriptions by prominent representatives of the field . . . tell us what their authors thought they were doing or what they wanted others to think they were doing, but they do not necessarily tell us what they actually did. The significance of such statements always needs to be assessed against independent evidence about actual practice.[41]

The assumption that actual researchers' behavior can be known by studying methodological rules is certainly not tenable. The difference between norm and deed cannot be ignored with impunity by anyone who wants to understand the deed. However, just as the interest of historical changes in *law,* for example, is not dependent on the question of whether or not history of law is the best route to becoming familiar with the actual behavior of people in a particular society, the importance of the history of methodology does not necessarily depend on whether or not actual research practices can be read from it. It is also true for both the history of science and the history of society that it would be naive simply to explain actual developments on the basis of given methodological or judicial precepts, but this intention does not have to be at the base of a historical investigation of changed norms and rules. Signaling that methodological styles may vary in different cultures and different times, and answering the question of why this is so, is of other interest.

Of course, the core question underlying this project is what difference would it make if researchers in fact *would* live up to the rules. If they did, would research then be the extracultural affair it is supposed to be? Or is it the other way around: If science and culture are not separated, is this due to the frequent violation of the rules? Would more control or positive and negative sanctions help? If not, why not?

Besides, methodological standards form the *conscience* of a scientific field. Students are trained within a particular methodological style; by studying manuals on methodology and "style-related" introductions to the discipline, through numerous separate methodological pointers and remarks, and by learning from exemplary models, assumptions and working methods become so much a matter of course that after a certain time other ways are unthinkingly perceived as incorrect. In other words, methodology is internalized. Furthermore, there are sanctions against transgression of the rules. Anyone who goes beyond the bounds must do that imperceptibly. To gain the appreciation of colleagues and to be able to publish in the journals recognized by them, it is necessary at the very least to be able to present the research in the prescribed forms in the retrospective account of it. Today, research *proposals* must also be drawn up according to codified standard regulations if they want to be eligible for subsidies. In brief, anyone who wants to acquire scientific status, or even just wants to survive in the scientific community, cannot ignore the rules of the game with impunity.

In these respects, too, the comparison with the rules of the law applies, or to vary the example, with the commandments of the church. Whereas the actual behavior of, let us say, Italian Catholics in the 1950s can be better studied in some other way than by perusing the Ten Commandments and the Papal Encyclicals, these texts did form their conscience and visible (or confessed) transgressions could be punished on the basis of these precepts. In this way the rules defined the margins of acceptable behavior. In the same way methodological criteria for scientific researchers do not determine and therefore describe behavior, but they do regulate it. They form the context within which variation of research behavior is sanctioned.

So, the rules of the game are not without consequence. In the first place – and that is my *first characteristic of methodology* – the rules define the professional field of a scientific discipline. In the accumulation of the number of "styles of reasoning" in the history of science the philosopher of science Ian Hacking even sees a form of scientific growth.[42] Every new style of reasoning regarding the correctness or incorrectness of statements, he argues, brings with it a new range of statements that can be researched. Statistical statements, for example, could be eligible only for the categories true and untrue after the rise of the statistical style of reasoning. Interpretations or value judgments, to add another example, can be dealt with only if there are agreements according to which *their* incorrectness or invalidity can be determined. Therefore methodological styles constitute new fields. With their accumulation in the course of history, according to Hacking, there is a growth in statements that are eligible to have their truth value determined.

There is another side to this: *Restriction* to a particular style limits the range of permissible questions and ideas. In scientific communities, where the definition of science is linked to quantitative ways of empirical testing, subjects that cannot be researched according to these criteria are regarded as not belonging to the scientific field. Those submitting proposals or articles that are therefore classified by the scientific community under other sectors such as "literature," "ethics," or "journalism" will not be very successful in their request for subsidy or in getting their articles published. On the other hand, in communities where value judgments are considered important or where the development of theory is considered the most important characteristic of science, a purely empirical-experimental proposal has little chance. In this sense methodological styles restrict the field of activity of a discipline.

In addition, methodology – and this is my *second characteristic* – affects the nature of the knowledge that a science produces. Methodological rules are not completely neutral instruments for distinguishing sound and nonsound statements and theories. In order to become eligible for treatment according to the rules, ideas must be given shape or translated into the appropriate terminology. In this way methodology is of consequence for the contents of the theories and statements produced.

An example that recurs more often in this book is psychological research on the relationship between personality and variables such as suitability for a profession or sensitivity to particular bodily illnesses. The introduction of predictive empirical-analytic methodology into Dutch psychology of the 1950s required that descriptions of character could be "operationalized" in terms of visible behavior for the sake of testability. A subject's underlying "real personality" could therefore no longer be a legitimate object of psychological investigation, but only his behavior, split into carefully defined variables. According to the new rules, a psychologist's inference about the existence of covert "hidden aggression" behind manifest "submissiveness" could no longer be included in psychological reports. These reports could then no longer be a "picture in words" as the *geisteswissenschaftliche* psychologists liked to call them. Well-known examples also include the numerous translations of psychoanalysis into behaviorist or intentional terms as accompanied by logical-positivist and hermeneutic methods of research respectively. Inevitably, after such a translation it is possible to see the translation process in the research results produced. Methodological rules, it can be seen from this, are not neutral but are significant for the content of the scientific product.[43]

Since researchers of the natural sciences arrived at the finding that its practitioners deviate on crucial points from the norms of the general philosophy of science, the philosophical and historical interest in methodology has not only been criticized, but the relatively strong orientation of the *social* sciences to rules and regulations in carrying out research has also been under attack.[44] These sciences, say the critics, with their preoccupations about being scientific, are trying to match qualities of an admired elder brother which this brother does not actually have at all. Representative of this is the viewpoint expressed by the sociologist of science, Steve Woolgar:

The debate about the SCIENTIFIC character of social sciences recurs from time to time – for example, in the German dispute about positivism – and is echoed in the tired old question found in almost all elementary introductions to sociology: Is sociology SCIENTIFIC? (the capitals denote the mythic, idealist connotations of this use of the term). . . . Perhaps the most significant achievement of the social study of science is the finding that the natural sciences themselves only rarely live up to the ideals of SCIENCE. . . . By recognizing the un-SCIENTIFIC character of both social and natural sciences, social scientists can stop worrying about how SCIENTIFIC they are.[45]

Insofar as Woolgar's criticism opposes the tendency to ascribe universal and timeless validity to one particular – for example, logical-positivist – idea of science, it is in my opinion correct (this point will be dealt with later). At the same time, however, it must be noted that this is a case of the kettle calling the pot black. No less than the objects of their critique, the critics elevate putative or real mores and habits of the natural sciences to a norm. If methodology in fact has little to contribute in the natural sciences, then that will also be the case in the social sciences. In spite of science studies' characteristic aversion to unjustified universal statements about science, as far as methodology is concerned, it is assumed without any further research that what is true for the natural sciences is true for the empirical sciences in general.

Whereas methodology is presented by philosophers of science as a system of rules for choosing among competing theories and whereas many in the field of science studies *deny* it a significant role in this respect, the rules of the social sciences quite often are not actually *intended* for making decisions on the theoretical level. Methodological manuals in the empirical-analytical style, for example, offer for the most part detailed rules and techniques for answering particular types of empirical questions, or for testing hypotheses not theoretically embedded, which at the most contain conjectures about empirical laws. As I will show in this book, for many of the other methodological styles it is also the case that the primary aim of their rules is not to guide the formation of theory into the proper channels or to regulate the choice between competing theories.[46] In the case of the social sciences, therefore, accounts of methodology of science studies often seem to be not much more empirically adequate than those of philosophy of science.

What is being overlooked is the fact that methodology in the social sciences has – to a greater degree than that of the natural sciences – some *extra functions* to perform. Social scientific products, more so than those of the natural sciences, are also made in other sectors of society. The state, journalism, literature, philosophy, and the church, among others, make claims and develop theories about social developments, individual identity, the operation of memory, the best ways of raising children or containing conflicts. The task that is ascribed to the social sciences, or that they claim to fulfill, is to provide knowledge of a special quality about such questions. To a lesser degree than with the products of the natural sciences, this quality can be seen in the products themselves. Social scientific texts are usually formulated in an idiom, which hardly differs at all from those of the nonscientific sectors named. The *difference* must be guaranteed and indicated by being able to refer to the way in which they were created. This fact alone means that both at home and abroad, much attention is devoted by social scientists to the manner of production. It is mainly the methodological rules that lend scientific identity to the social sciences. By referring to the "how" of the discipline, these sciences demarcate themselves from their many nonscientific competitors and so legitimatize their place in society. This legitimating function is the *third* quality of social scientific methodology that I want to point out.[47]

There is yet another respect in which social scientific methodology differs to some degree from that of the natural sciences. The rules of the social sciences are not merely means of production within their own disciplines; they themselves function as a product of these disciplines. In an article with the illustrative title, "The Social Scientist as Methodological Servant of the Experimenting Society," the American psychologist Donald Campbell even argued that as far as their production is concerned these sciences (should) *mainly* depend on their methodology: "The aspect of the social sciences that is to be applied is primarily its research methodology rather than its descriptive theory."[48] The sociologist who disputes on methodological grounds the police's claim that crime figures have risen, the historian who in a debate on "collaborating writers in wartime" raises presentist approaches to the past for discussion, and the psychologist who operates as a scientific advisor in a government committee for research into alternative medicine are all intervening in debates outside their own discipline. More important, they do not enter into these debates with empirical or theoretical knowledge, but with the methods and techniques of their discipline. This *fourth characteristic* of social-scientific methodology, the characteristic that methodology is a form of output, is also related to the fact that people and their mutual relationships are the object of consideration in numerous other social sectors. In this way, methods and techniques of social-scientific research, intended in the first instance for colleagues, also find a market as aids in the questions, plans, or decision making of third parties.

Together, these four characteristics of methodology form an argument for the importance of research on methodology as an independent aspect of the social sciences. They also lead to conjectures about the way in which methodological variability can be understood.

Changing methodology

How does a scientific community come to create or subscribe to a particular methodological style? Or rather, what is the process whereby a group of scientific practitioners assumes a new scientific identity with a new methodological style?

One option is that old rules simply turned out not to be *adequate* and that the new ones were more suitable for the questions that a discipline or group concerns itself with. The interpretative approach to applicants in prewar Dutch selection psychology would then have been pushed into the background because applicants can be assessed better with the empirical-analytical methods that replaced them. It is not *necessarily* by elevating contemporary ideas to a norm in a presentist fashion that a historical analysis may lead to such a conclusion. However, if a rejected methodological style was once subscribed to by trained and educated practitioners of the discipline over a substantial period of time and if these practitioners also scored social successes, then during all that time the psychologists involved must have shared a collective lack of insight into the correct approach and their clientele must have accepted an unusable product for years. Methodological diversity in the same period would indicate great cognitive confusion within the discipline. Although on the one hand something like that cannot be automatically excluded, on the other hand it is not such a plausible option that it would be the *first* to be eligible for further study in concrete cases.

If we bear in mind the four characteristics of methodology I have discussed, then other possibilities present themselves. First, the fact that methodological rules restrict the field to be studied would appear to offer some lead. Certain questions to be answered and problems to be solved could have led to particular methodological rules. Looked at in this way, *new* questions and problems that in the course of time a discipline came to appropriate would form the actual background of new research methods, whether derived from others or developed by the discipline itself. The objection to this view of affairs, though, can be made that it is not precise enough. When the rise of particular methods is traced back to the rise of particular questions or research topics, it is implicitly assumed that the methods are dictated by the questions. However, reality does not itself prescribe how it is to be studied. There is a certain latitude or play between the questions addressed in science and the methods employed to answer them. The retrieval of the original questions that prompted the development of a methodological style is therefore not sufficient to clarify why a specific system of prescriptions became the general standard in a particular period or for a particular group of investigators. Even if the original research topics of a discipline are traced back, the question remains open as to why these particular methods were chosen and not others.

If reality does not dictate methodology, researchers will have to work it out or choose it themselves. In other words, the existence of this latitude in the relation between the questions and the methods of a field also implies that methodological matters are *debatable*. And that raises the question of who the *participants* are in the debate on methodology. It can be argued that this group will not be confined to the scientific practitioners themselves, but that their "clientele" also have an input here.[49] To begin with, this point can be argued on the basis of the *second* characteristic of methodology previously mentioned, that is, methodological rules are not neutral with respect to the questions that are handled by the discipline. A particular methodology is accompanied by a particular formation or transformation of questions, in such a way that these questions can be dealt with according to the rules. For instance, the fact that the introduction of the predictive empirical-analytical methodology into Dutch psychology of the 1950s implied that a subject's underlying "real personality" could no longer be a legitimate object of psychological investigation and that statements concerning the essential aspects of a person's character were banned from the psychological reports.

Such a methodology-steered redefinition of the question must be acceptable to the discipline's clientele. Or the other way round: New interpretations of problems can lead to a situation in which the usual methodology is no longer adequate. In other words, to allow changes of methodological style to "go through," it must be possible, socially speaking, to create the space for it, or to have a preexisting "gap." Even though scientific practitioners can certainly come up with methodological solutions themselves and even though they also undeniably derive insights from other sciences or philosophical traditions in their own countries or abroad, insofar as these solutions are not acceptable to their environment, they will have to amend them, at the very least. In the "space" between the questions and the methods a fine tuning of perspectives takes place between the practitioners and the clientele of the discipline. So when a new methodological style has come into vogue, it is not only the professionals who committed the rules to writing who will have had a vote in methodological matters, but the clientele of the discipline as well.

The relevant clientele of a discipline can also be expected to have a say because of the legitimating function of social science methodology. As I pointed out, the insights of the social sciences quite often are put in the same idiom as that of their many nonscientific competitors and the difference must be indicated by reference to the methodologically guaranteed ways in which these insights were obtained. Methodology provides the social sciences with their scientific identity. Of course this can be done successfully only if the rules appear *trustworthy* to those groups to which a particular science has to legitimate its existence. Nowadays in the Netherlands an appeal to the "special empathy" of an assessment psychologist is no longer a convincing argument for the validity of his or her judgments, but such a legitimation was acceptable in the 1930s (see Chapter 2).

Finally, a reason to expect that the rules of the social sciences will be open for discussion with these sciences' clientele is to be found in the fourth feature of methodology, that is, that the rules function as scientific output on their own. Naturally, this too is possible only if the rules are acceptable to those who are supposed to accept them. If the prospective clientele is not convinced by the methods and techniques, they will reject them as inadequate.

In short, whereas earlier I concluded that methodological rules cannot be dictated by reality, methodologists cannot dictate rules either. In order to determine why a particular methodological style was subscribed to or abandoned in a particular period or scientific group, it is necessary to find out which social groupings were a discipline's most important clientele and what their *premises* were with regard to the problems to be dealt with and the solutions expected. Therefore I will investigate to which social groupings (in governmental bodies, churches and nonchurch institutions, companies, and universities, for example) the proponents of the various methodological styles were directing their arguments and ideas and what these groups thought about the questions that were raised for discussion.

In other words, it is fruitful to assume that in a methodological style, apart from the *scientific* identity of the discipline, its *social identity* is given expression. A relatively strong unanimity on a new methodological style, that is, a *change* of scientific identity, will therefore also point to a change of social identity. Studying the process in which a discipline or a group disposes of an old scientific identity and assumes a new one demands the discovery of the changing social identity of that discipline or group.

This leads to a different image than that of the cumulative insight into contemporary standards for science. Just as in histories that describe "the influence of external factors," in this book I will look for the ways in which nonpsychologists participated in the discussion. Nevertheless, there are differences between my way of working and the usual "externalistic" histories, too. I would designate only my approach as being externalistic or rather, contextualistic, with a dual qualification.

First, my account distances itself even further from the standard image than externalistic interpretations generally do. Historians who try to refute the image of an internal cumulative progression often contribute to the idea of external factors by explaining developments on the basis of the origin of the *questions* confronting a discipline. If those questions originate in groups other than that of the scientific practitioners themselves, then scientific thought and action are rooted in *social* needs, it is argued, and therefore the development of knowledge is said to be not purely scientific but a sociopolitical affair. However, for years now the image of specific social developments that have led to specific tasks for the social sciences and with them to the rise and establishment of particular methods and knowledge has in fact been part of the *standard* histories. The historical introductions to basic texts in test psychology, for example, invariably note the large significance of the First World War for the development of the psychological test in the United States (many recruits had to be selected in a short period of time). These texts can do this without being self-contradictory because in fact such a role hardly tampers with the idea of an autonomous development of science. After all, this type of example shows only that certain social conditions had to be met before "the scientific approach" had the chance to establish itself. Society has the role of merely being a provider of questions; it has no say in "the real scientific matters." This sort of history is therefore far less a refutation of the standard image than externalists may believe it to be.

As I explained earlier, I do not follow such an approach. I argued that the questions as such cannot determine the methods and that in the latitude between questions and methods there must be a discussion between the psychologists and their clientele. In my view, nonpsychologists also have a say in a domain that is closed to them in the picture that portrays them merely as producers of questions. My second qualification is that I will not apply any definition of which factors must be labeled "internal" and which "external," as is usually done implicitly in externalist or contextualist accounts. If definitions are applied, as I argued before, an ahistorical idea of science is still elevated to a norm and science and environment are still separated from each other. Conversely, in the following chapters the process is investigated whereby the boundaries of psychology, as part of a society that is both multiform and changing, and in a dialogue between psychologists and their environment, are defined and redefined.

Insofar as this will appear later on to have yielded a convincing picture, the conclusion will be justified that the methodological variability of Dutch psychology must not be ascribed to a slow or steady growth of insight into the correct approach, and that the former deviations from today's standards should not be ascribed to intellectual confusion in the past. The alternative image will be that psychology in the Netherlands affiliated itself to many different social groups and that a sociocultural pluralism is expressed in its varied methodology. In each subsequent chapter the history of Dutch psychology will be entered through a different gateway. In Chapter 5 the various lines of the book will come together.

2

Handwriting and character

In 1948 the Dutch Psychological Association[1] invited a number of graphologists to deliver lectures during its "Psychologists' Day" on the interpretation of character from handwriting. The reason for this was "that popular interest and practical application . . . means that it is desirable to bring graphology before the forum, in order to examine a number of aspects of it."[2] Their reception cannot have been very cordial. "The host called his guests riffraff and cranks and let them know what an honor it was to be received by him" is how one of these graphologists would later describe the atmosphere.[3]

At the time graphologists had little to fear from such critical psychologists, though. In the early 1950s, 75 percent of the psychological consultancies and psychology departments of companies were using graphology.[4] Companies where graphologists worked included, for example, the National Railways, the Post Office, the Royal Dutch Airlines, the Heineken Brewery, and Philips. Handwriting was also interpreted at the National Psychological Department and the (private) Dutch Foundation for Psychotechnics in Utrecht. Graphologists were employed in the Laboratory for Applied Psychology at the Amsterdam Free University and in 1953, the University of Leiden even appointed an "Extraordinary Professor"[5] of Graphology,[6] C. J. F. Böttcher (also professor of physical chemistry), who had already been teaching graphology as a compulsory subject to psychology students in Leiden since 1949. Beginning in 1947, students of psychology at the University of Utrecht had to choose between a compulsory course in graphology or Rorschach.

Some parts of this chapter have previously been published in English in T. Dehue, "Why Methodology Changes: Transforming Psychology in The Netherlands I," *History of the Human Sciences* 4 (1991):335–51.

Nevertheless, in retrospect it would become clear that the Psychologists' Day of 1948, with its tense atmosphere, was more than an isolated incident. In the 1950s and 60s relations between psychology and graphology worsened. Psychologists bent on innovation, who at that time were making a case for an empirical-analytical approach in psychological diagnostics, had identified graphology as a prototype of the *geisteswissenschaftliche* interpretative methods that they had declared war on, and used it as a negative frame of reference. In the 1960s the heated debates between proponents and opponents of graphology even shifted from the respective professional journals, the *Dutch Journal of Graphology* (*Nederlands Tijdschrift voor de Grafologie*, hereafter *NTG*) and the *Dutch Journal of Psychology* (*Nederlands Tijdschrift voor de Psychologie*, or *NTP*), to the national press. Nearly all large and small newspapers and weeklies threw open their columns to the contending parties.[7]

Graphology lost this battle. By now, it has almost completely lost its market in the field of personnel selection. Unlike in France or Germany, for example,[8] it is rare nowadays in the Netherlands to see personnel advertisements in which handwritten letters of application are required. In 1980 the Dutch Association for Personnel Policy even prohibited this requirement in its Application Code.[9]

In this chapter the different "frames of thought" of interpretative and empirical-analytical psychological diagnostics in the history of Dutch psychology are illuminated with the help of this controversy around graphology. As I argued in the first chapter, controversies offer a special opportunity to form clear images and to analyze how "sticks and pieces ended up as a ship in a bottle." Just as such a ship was once a pile of sticks and cloth, scientific insights do not already exist before they are discovered, they are constructed. In this context, the ship in the bottle is contemporary empirical-analytical methodology in Dutch test psychology. The question of how it got in the bottle is the question of how it has come about that today this empirical-analytical methodology is more or less generally regarded by Dutch psychologists as being more rational than that of the interpretative approach.

Answering that question is the ultimate objective of this chapter. In the final sections it is split up into two partial questions: First, how was it possible that the conventions of interpretative diagnostics, which *now* sound so unconvincing to most psychologists, were once able to seduce people and institutions into spending money and effort on psychological diagnostics? Second, following from the first question, how could these conditions of existence for an interpretative psychology of assessment *disappear*? In other words, what led to its loss of persuasiveness and to the acceptance of empirical-analytical methodology? That these new criteria are more rational than the old ones is therefore not regarded as the *source* of the changes, but as a product of history.

Empirical-analytical and interpretative psychodiagnostics

The differences between forms of empirical-analytical and interpretative thinking do not date merely from the postwar period. From almost the earliest days of psychology as an applied discipline in the Netherlands, differences can be found in the methodological stands and working methods of psychologists. For a proper understanding of the developments we have to begin at the end of the previous century.

As in other countries, the practice of academic psychology in the Netherlands originally amounted chiefly to the investigation of human consciousness, not – as traditionally regarded – with philosophical methods, but with experimental methods derived from the natural sciences. The most important Dutch representative of this empirical-analytical approach to psychological matters was the philosopher Gerard Heymans, who in 1890 became professor of "The History of Philosophy, Logic, Metaphysics, and The Science of The Soul ['*Zielkunde*']"[10] in the northern city of Groningen. Heymans distinguished himself from his Dutch colleagues, whose teaching commitment also usually included "science of the soul," first by giving the discipline an important place in his teaching and research, and second, by interpreting it as an empirical discipline rather than as a philosophical one.

Heymans had earlier analyzed "the laws and elements of scientific thought" for the natural sciences.[11] Just like Theodor Fechner, whom he criticized but who nevertheless influenced him profoundly,[12] Heymans was convinced that these laws of scientific thinking also applied to psychology:

> We have not a single reason to doubt that the impression produced by the melody as a whole, is as much a sum-effect of causes, as the movement of the planets is a sum-effect of inertia and gravitation. The same view holds as to the concurrence of different mental contents in inhibition, of different motives in willing, of different arguments in producing beliefs, of different tendencies and predispositions in the formation of character, and in many other cases.[13]

In psychological experiments, the relation between two experimentally produced variables should be examined systematically in order to arrive, through induction, at the causal laws of human consciousness. Where experimental manipulation of variables is impossible, correlational research into the merging of various relations that exist in reality should be carried out. In 1892 Heymans founded the first laboratory for experimental psychology in the Netherlands.[14]

Heymans was also one of the first empirically oriented psychologists who carried out research in this fashion on the individual *differences* between people. In this, too, he was concerned with tracking down the laws that lay behind those differences. He interpreted individuality as a specific combination of a number of general human characteristics.[15]

Around 1920, possibilities also began to arise in the field of vocational guidance and personnel selection. Although this developed a few decades later in the Netherlands than in surrounding countries, the mechanization and rationalization of industry with its accompanying differentiation of labor was now well under way. New kinds of jobs came into being, such as the typist, telephonist, tramdriver, high-voltage electrician, and aviator. In private companies before the 1920s, the son followed his father's profession as a matter of course, or tradition and background otherwise determined how people would earn their living. In industrialized society with its newly created jobs that was no longer possible. So, the equally new profession of assessment specialist for personnel selection and vocational guidance came into being.

The psychologists who had developed apparatus in their laboratories for measuring human motor and sensory systems were able to propose with some success the idea that they were specially capable of assessing individual capacities. Like teachers, engineers, theologians, doctors, and others who practiced "psychotechnics,"[16] they embarked on their new task.

Henri J. F. W. Brugmans, a student and later Heymans's successor, was one of the first in the Netherlands who operated as an academically trained psychologist outside the university. In 1920 the Dr. D. Bos Trust had been founded "to undertake or support socio-pedagogical work to promote free people's development in the broadest sense of the word." To achieve this goal the trust had called on Brugmans for assistance. Brugmans was of the opinion that the "free people's development" could best be served by concentrating on research on individuals' suitability for education or a profession. Brugmans was appointed director of the newly founded Dr. D. Bos *Foundation* and together with his assistant, Jacob L. Prak, he set to work.

With the help of what they had learned from Heymans and in imitation of the German-American psychologist Hugo Münsterberg, they constructed series of tests for the selection of, among others, telephone operators and high-voltage electricians. In the "Announcements of the Dr. D. Bos Foundation" (*Mededeelingen van de Dr. D. Bosstichting*), their plan was set out in detail. The psychologist made a "function analysis" of the sensorimotor and cognitive capacities required for a particular profession. He then carried out a series of experiments and tests to measure these capacities. For example, the "psychological analysis" of the telephonist at the Dutch Post Office concluded that for this work the capacities of "concentration," "attentiveness," "auditory and visual memory," "steadiness of hand," "rapidity of arm movement," "strong disposition to reactivity," and "relative passivity" were required. Except for the two last characteristics, which according to Brugmans could be guaranteed by having this profession staffed with women, tests were designed for all of these traits. The measurement of concentration took place by having the telephonist simultaneously write out the alphabet and recite the series of uneven numbers for a minute. To measure steadiness of hand the internationally renowned "tremometer" was applied: This involved a copper plate in which figures were cut out, which then had to be traced with a metal stylus without touching the edges. If that happened a bell rang.[17]

Meanwhile, the company or institution that requested the psychotechnical advice had selected a series of people who had already been working for some time in the relevant profession and had drawn up a "variation series" of good-to-less-good workers. These workers then went through the whole series of tests and afterward the correlation between the variation series obtained in this way and those of the management were calculated. Tests in which this correlation was high enough were kept, whereas others were removed (provided that, according to the psychologists, the management had made no mistakes in judgment). Thus a pool was built up of psychotechnic means for determining the suitability of later applicants and those asking for vocational advice.

Brugmans and Prak were successful in this approach, at least to the extent that in 1924 Prak was the first Dutch psychologist to be employed by a company (Philips). One of his tasks at Philips was to select assistants for the physics laboratory. To compose a series of tests Prak studied American test literature; the Army Alpha and the Army Beta, selection tests for the American army, were a source of inspiration. The Jesuit Jacques van Ginneken was working from a similar viewpoint in the "Psychological Vocations Office" (*Zielkundig Beroepskantoor*), founded in 1919 in Utrecht. The same was true of the (female) physicist and psychologist R. A. Biegel, who in 1926 set up a "psychotechnical laboratory" in the Dutch Post Office at The Hague.

Whereas these kinds of procedures soon became established in the United States and England,[18] other countries such as Germany[19] and the Netherlands were quick to raise objections. In the "elementaristic tests" a subject was compared to a control group of the same sort and he or she was regarded as a conglomerate of separate capacities. However, it was argued that no one can be regarded as a specimen of a species and every individual is, literally, an indivisible whole. Furthermore, there was also a feeling that thinking in terms of laws, which this approach was based on, deprives people of their freedom of choice and responsibility. For these reasons, on the tenth anniversary of the Amsterdam Vocations Office (*Amsterdams Beroepskantoor*) in 1930, the Dutchman F. J. M. A. Roels, professor of psychology at the University of Utrecht, argued passionately for the introduction of a new psychotechnics inspired by the German *geisteswissenschaftliche Philosophie:* "The time of the analytical method . . . is gone. . . . If one dissects man into pieces, one should not be surprised if the main thing of man, life, has disappeared. It is the same in the psychological field, if one breaks the personality into pieces, then the blood, the life, is gone."[20]

Therefore instead of regarding a subject as a "thing with functions," the subject must be regarded as a *person* with his own *character*. In this sense the concept of "character" stood for an underlying reality of which visible behavior is only one of the possible manifestations. The psychologist should be interested in what someone is "essentially" and not only in the superficial exterior forms. According to this approach, the "feeling" and "the will" especially, which are not covered by the analytical approach, should be studied.[21]

The ways in which these deeper layers of the personality were exposed differed considerably from the ways in which empirical-analytical psychology studied human capacities. The methodological basic principle of this interpretative psychology was that people compile information about their fellow humans in a different way than they do about inanimate nature. People are capable of understanding the significance of other people's acts and statements. In doing so they make use of their intuitive ability. This ability can be sharpened by good training and experience. In order to indicate the difference between the trained and everyday understanding, the word *verstehen,* borrowed from the German philosopher Wilhelm Dilthey, was used in the Netherlands. For the purpose of this *verstehen,* specific methods were used to elicit actions and expressions from the person to be assessed. The psychologist could use these actions and expressions in forming an image.

The methods of the Dutch Foundation for Psychotechnics in Utrecht, founded in 1927 by the theologian and later psychologist David J. van Lennep (together with the teacher of classical languages Tako Kuiper) offer a good example of the new mode of thought about personnel selection. To begin with, in its early years the Foundation recruited no academic psychologist for its own staff. Not only were academic psycholo-

Figure 2. From D. J. van Lennep, *Psychotechniek als Kompas voor het Beroep* (Utrecht: De Haan, 1949), p. 65.

gists considered to have the wrong training, but the Foundation also preferred staff members who had experience in the professions for which applicants had to be selected. Such staff members were able to immerse themselves in the task of these professions, it was argued. Training in the use of psychological methods was provided by the Foundation for Psychotechnics itself.

It was characteristic of these methods that subjects were not forced to give a specific response to a question or a task, but were instead encouraged to give self-disclosure," as Van Lennep called it. He also labeled these methods as "multivalent" or "polyvalent" tests because the applicant's reactions and answers were not sorted out in terms of right and wrong, but were interpreted characterologically.[22]

Among the methods for acquiring material for interpretation were the so-called observation tests. These tests included, for example, the performance of a task by the applicants, not with the aim of investigating whether they could complete this task, but to discover their personality in the way they approached it.

The "physiognomic" methods, where the applicants' appearance was interpreted, also belonged to the observation tests. Van Lennep always emphasized that the research subject must be a whole and not a compilation of separate parts. Physiognomy, for example, which based itself on general rules such as "a strongly projecting chin indicates willpower," was rejected. The idea that physiognomic interpretation could not be "a statistical affair," but required a "sensitive and trained eye" because differences in detail in an appearance influence the entire appearance was illustrated by Van Lennep using drawings like that of the faces reproduced in Figure 2.[23] "The impassioned sensuality in the expression of the drawing in the middle," he explained, "is reduced if the mouth (see drawing on the left) or the eyes (see drawing on the right) are drawn smaller. The smaller mouth diminishes the expression of sensuality, and the smaller eyes reduce the expression of passion." He added that, however, "the field of observation of the physiognomist-psychologist" does not end "where the collar of his test person begins."[24] It is only *in drawing* someone's appearance, Van Lennep continued, that you really perceive its proportions. This also increases the researcher's powers of empathy. A portraitist was therefore employed by the Foundation for Psychotechnics (see his drawing of the psychologist De Groot in Figure 5 later in this chapter).

The most important category of self-revealing or polyvalent tests, however, was that of the "projection tests." These were based on the idea, familiar from psychoanalysis, that in ascribing qualities to other people or things, people unwittingly reveal various things about their own unconscious goals and motives. In the 1930s Van Lennep himself designed and painted the *Four Pictures Test,* whereby the subject had to come up with a coherent story for four vaguely colored pictures.[25] Some years later the Rorschach test was also included in the Foundation's repertoire. In this well-known test the subject is asked to tell what he or she sees in a series of ink blots.

The task of the psychologist was to interpret the resulting stories. Generally, he was not completely free in his interpretations. The tests were standardized in the sense that in the instructions, categories were given into which the reactions had to be divided according to particular rules. For example, with the Rorschach every answer, to begin with, had to be classified according to "localization" (the part of the blot on which the projection of the subject is concentrated), the "determinant" (the aspect of the spot – form, color – that led to the interpretation), and the "content" (the nature of what is perceived in the blot: a person, animal, sort of material, etc.). These main categories were then subdivided again. Scoring the Rorschach reactions was a rather laborious chore.

Finally, in writing the report the psychologist should not present figures but be "a painter with words": "As in a real painting in which the painter adds, through small accents, extra light or sections in shadow, such contrasting reliefs to the picture that the viewer has no difficulty in understanding the interpretation of the painter, so the editor must give a clear interpretation of what has emerged from the psychological study," Van Lennep told the staff members of his Foundation.[26]

Unlike tests in the experimental school, which by this time had been rapidly pushed into the background, the tests from the interpretative tradition were not correlationally checked on an experimental group before being applied. Comparing someone's score to a norm constructed on the basis of the score of others did not tally with the principles of uniqueness and freedom of each individual. The correctness of an interpretation had to be proved in other ways. Van Lennep repeated what the German phenomenologist and psychiatrist Karl Jaspers had said about the "hermeneutic circle" applied to interpretation: Every expression can be understood only on the basis of the entirety of expressions

Figure 3. One of the pictures of Van Lennep's Four Pictures Test (second edition. Utrecht: Nederlandse Stichting voor Psychotechniek, 1958)

and the entirety can be known only through the separate expressions. You have to jump into the circle somewhere and then continue to confront part and whole with each other until a "feeling of evidentness" comes into being about the validity of the interpretation. For the rest, "verification" was carried out afterward by checking the extent that given selection or career choice recommendations were followed and by asking whether or not the recommendations followed, in retrospect, were seen as the right ones.

Graphology and interpretative psychodiagnostics

Graphology is older than academic psychology and to a great extent developed separately from it.[27] However, when practice-oriented psychology began to make headway at the beginning of the twentieth century, Dutch graphologists also began to seek scientific status, and to this end attempted to link up with psychology. The psychologists from the empirical-analytical tradition were dubious. Interpretration of handwriting was regarded as pseudo-science and charlatanism.[28] However, with the success of interpretative psychodiagnostics, a connection did come into being between psychology and graphology. The principles of graphology were very much the same as those of interpretative psychodiagnostics.

There is a broad range of introductory works that shed light on the graphologist's manner of thinking and operating. The book, *The Scientific Foundations of Graphology* (*De Wetenschappelijke Grondslagen van de Grafologie*), written in 1949 by the Dutchman S. V. Margadant, is representative of graphology in the Netherlands.[29] Margadant was a member of the the the Dutch Society for the Enhancement of Scientific Graphology (*Nederlandse Vereniging tot Bevordering der Wetenschappelijke Grafologie*) and editor of the *Dutch Journal of Graphology* (*Nederlands Tijdschrift voor de Grafologie*). Furthermore, he based his work on that of the famous German characterologist and graphologist Ludwig Klages, author of *Handschrift und Charakter* (1917), which had already been reprinted twenty-two times by the time Margadant's book appeared,[30] and on other canonical works of international graphology.

The basic principle of the interpretation of handwriting, Margadant wrote, is that people's character is expressed in the totality of their unconscious movements. And because handwriting is the permanent trace of a large number of unconscious movements, personality is expressed in it. This trace must be followed with "a loupe which is not too strong" or with "a stick with a point which is not too sharp." While tracing someone's handwriting, the graphologist experiences the movements, perceptions, and feelings of the writer.

Separate characteristics of handwriting, such as small writing, must be examined in the context of the entire writing. Someone who writes small can be anxious, or shy, but also narrow-minded or a "man for the finer point." A graphologist finds out exactly which quality is involved by looking at the *milieu graphique* (term derived from J. Crepieux-Jamin, *L'Ecriture et le Caractère,* first published around 1885). In addition, the "basic rhythm" of the writing must be watched. Margadant derived that concept from the German graphologist R. Wieser (author of *Der Rhythmus in der Verbrecherhandschrift,* 1938). By examining the whole writing letter by letter, the graphologist can ascertain whether the writing is "elastic, rigid, weak, or disturbed." Depending on this it can be determined whether a positive or negative pole of a quality is involved. The basic rhythm therefore determines whether someone's equanimity arises from "depth, tranquility, and reflection" or from "sober common sense caused by a cold heart."

The symbolism of the spatial opposites "above and below" and "left and right" is also important. The characteristic aspect of the above–below symbolism is a higher appreciation of what is above. A writer unconsciously perceives the upper zone of the handwriting as a spiritual activity, the middle zone as that of conscious acts in daily life, and the lower zone as that of the instincts and the life of the emotions and drives. Right and left represent the spatial goal-directed forward–backward symbolism, together with the temporal former–later or past–future symbolism. These facts on the left–right symbolism and the above–below symbolism must be interpreted together by the graphologist.

In this way, it was thought, the essence of the personality in particular can be discovered. Following on from Klages, Margadant emphasized that if the results of a carefully executed study turn out to contradict a judgment obtained in other ways, it should not be concluded that the graphological interpretation was not correct, but that it revealed the unconscious characteristics and the essence behind the visible behavior.

The interest in the essence of the person, the empathizing manner of acquiring knowledge, the idea that people are unconsciously revealed in their expressions, the idea that no single aspect should be studied independently of other ones, that rather the personality as a whole should be studied: These were all characteristics of both graphology and interpretative psychodiagnostics. Because of the large degree of similarity between the two fields and also because graphologists themselves wanted nothing more than recognition by academic psychology, the way was open for psychologists to include graphology in their package of diagnostic resources after the swing to the interpretative mode had taken place. "In any case, graphology is a method which, basically spoken, fits in completely with the whole of psychotechnical methods," wrote Van Lennep, who had also trained himself in graphology.[31] To facilitate the "inner empathy with the movement of writing," he even installed a so-called graphoscope in his institute as a kind of "empathy machine." While an applicant sat writing, the researcher could follow in enlarged form the genesis of the writing through a camera and screen, and through two tubes in the psychotechnician's ears he could "listen to the writing in its rhythmic process." If desired, it was even possible to record the sound on a gramophone disc. According to Van Lennep, willpower and perseverance especially, but also ambition, vanity, and egoism could be diagnosed in this way.[32]

Only the occasional individual, such as the America-oriented Prak,

Figure 4. The graphoscope in use (photo from the Archives of the History of Dutch Psychology, Groningen).

who in the meantime, incidentally, had expanded his surname by adding one of his first names, to Luning Prak,[33] still rejected graphology completely by this time. From his Psychological Institute The Hague (*Psychologisch Instituut Den Haag*) founded in 1935, he rejected "the uncritical and insufficiently scientific position" of graphology's practitioners, who refused to confine themselves to a limited number of well-defined and verifiable traits.[34]

However, most psychologists at this time only warned against graphology as an *autonomous* means of psychodiagnostics, operated by assessors who did not have any academic psychological training. Most psychologists were becoming proficient in graphology from the 1930s on, and they in fact set graphologists to work under scientific psychological guidance in their universities, municipal and private institutes, and in the psychology departments of companies.[35] At the beginning of the 1950s consultations were even under way between the Dutch Association of Practicing Psychologists (*Nederlands Instituut van Praktizerende Psychologen*, hereafter NIPP)[36] and the Dutch Society for the Enhancement of Scientific Graphology, with the aim of introducing an exam in psychology for graphologists. By passing this examination, the graphologists could acquire the status of "psychological assistant."[37]

That last plan was never put into practice, though. In the same period, psychodiagnosticians, who advocated a renewal and revival of the empirical-analytical approach, powerfully and successfully raised their objections to the interpretative methods. Considering that graphologists were traditionally seen as outsiders, it was not surprising that it was the interpretation of handwriting that became an important starting point for change: The most vulnerable spot of *geisteswissenschaft* methodology was the one to be hit first, and hardest.

On the one hand this outsider's position does lessen the representativeness of graphology as a case for studying interpretative thought in psychology (not all graphologists were academically trained and some were less capable or less interested in academic precision than their psychologist-protectors were), but on the other hand it is the *intensity* of the methodological debate associated with it that makes graphology such an illustrative example. The "graphology debate" clearly shows the intensive encounter of two different "language games" and the period of confusion that psychodiagnostics experienced in the first two postwar decades, as a consequence of it.

The verification of interpretative statements

Above all it was the psychologist De Groot, trained at the University of Amsterdam and employed by Luning Prak, who in the first postwar years was instrumental in starting the discussion of interpretative psychodiagnostics. That discussion began in 1946–7 with a "verifying study" of graphological statements, carried out from Luning Prak's institute. De Groot collected ascriptions of traits on the basis of graphologists' handwriting analyses and ascriptions of traits by people who knew the subjects who had provided the handwriting samples well. He then calculated the correlations between them. In 1947 he reported that the correlations had worked out very low.[38] Flabbergasted and indignant, Margadant reacted to the new criteria: "It is gradually becoming a game to check graphology. . . . Recently again someone felt compelled to put graphology through the mill, and to even publish the results."[39]

Such tests had little value in Margadant's view. He doubted the value of the criterion used. When his judgment is disputed, he said, one can be sure that the graphologist has only had *too much* insight. He then "touched on the essence" of the person, hidden behind his or her social and professional overt behavior. In his reply De Groot showed who had academic training and who had not: "Let Margadant be convinced that the distance between the contents of his article and real scientific thought is still particularly large."[40]

In the following years, first as assistant professor (1949) and then as full professor (1950) in applied psychology in Amsterdam, De Groot further developed his idea of "real scientific thought." In 1950 he gave a lecture entitled, "Towards a Crisis in Applied Psychology," in which he discussed the main features of the Continental-European approach.[41] In his view, these features were the creation of a contrast with the natural sciences-based thinking, an attempt to link up with great philosophers like Plato, Hegel, and Husserl, no connection at all with empirical-scientific methods, and language full of neologisms, complicated syntax, and circular reasoning: "The *Urdeutsch* sound of the words is . . . often more important than clarity."[42]

The direct object of this strong attack was once again graphology. At the request of the *Dutch Journal of Graphology,* De Groot had reviewed the new book by the Swiss graphologist M. Pulver, *Intelligenz im Schrift-ausdruck.* As he had written in that review, he considered the book representative of interpretative thought, which to his mind meant that it was "unreadable," had "a profound-obscure style," was full of "termino-logical vagueness and closed verbal concoctions," and was fit for "semi-intellectuals who take lack of clarity for profundity."[43]

Ten days later De Groot once again mounted the lectern. This time it was for a speech on the acceptance of his professorship in applied psychology.[44] Once again, but now in more circumspect wording suitable to the occasion, he mentioned his objections to the interpretative meth-ods. It was also in this speech that he first suggested combining method-ological elements from the Continental-European and Anglo-Saxon per-spectives (see Chapter 1). "There can be no objection raised to a methodological development . . . of our intuitive understanding of our fellow man as such," he argued, "but the assumption that the results achieved in this way do not need testing, that 'the essence' can really be 'witnessed' by sufficiently gifted and sufficiently trained individuals, without any possibility of objective criticism or testing, is a position in conflict with the basic principles of empirical science."[45] The existing interpretative research methods are admissible, he argued, but not for drawing conclusions. They can serve as tools for arriving at hypotheses with regard to an individual, from which verifiable *predictions* must be derived. Only in this way can psychologists arrive at objective empirical statements.

In this manner De Groot applied criteria from the empirical-analytical school to interpretative methods. Some aspects of interpretative thinking were retained. That is also clearly seen in an article published in English in 1954, in which he took a position against the German-born British psychologist Hans J. Eysenck. Eysenck defended the idea that the so-called unique individual is simply the point of intersection of a number of quantitative variables. De Groot argued against this by stating that it can make sense to consider an individual as unique. However, he also claimed that some criteria of science should be kept in mind. He presented his own "sketch of a scientific solution," which meant that a psychological report should be regarded as "a system of mutually related hypotheses with regard to individual behaviors, in both the past and future of the subject."[46]

Nevertheless, in this way the norms of interpretative thinking were for the most part transgressed. The requirement that character descriptions must be recorded in terms of the subject's observable activities for the sake of testability meant that only someone's *behavior,* and not the "personality" or the "essence behind the observable," can be the object of psychodiagnostic research. De Groot's idea also implied that the meaning of the concepts used would henceforth be precisely defined. Looked at in this way, a "painting with words," where the precise meaning of the terms is determined by the specific context in which they are used, could no longer be a scientific report. In addition, De Groot's emphasis on predicting behavior was perceived by many psychologists as an inadmissible denial of human *mutability.* "One does not predict what the subject will do, but indicates what should be done in order to make particular behavior more or less probable," wrote a fellow professor in applied psychology from Groningen University.[47] Above all however, the "phenomenological" psychologists at the University of Utrecht registered protests (see Chapter 3).

The confrontation of two styles in the "Graphology Working Group"

In 1954 the Dutch Society of Labor Psychology (*Nederlandse Vereniging voor Bedrijfspsychologie*) set up a working group that was supposed to advise on graphology as a psychodiagnostic instrument: first, on the proper use of graphology in the business sector; second, on the ethical rules that should be taken into account in graphological research; third, on psychological training for graphologists; and finally, on the stimulation of research on graphology.

Both psychologists and graphologists were appointed to the group, as well as some psychologists who had trained themselves in graphology or had appointed graphologists in their university institutes. De Groot belonged to the first category. Böttcher, the professor of physical chemis-

Figure 5. Drawing of De Groot by Stefan Stróbl, who was employed in Van Lennep's *Foundation for Psychotechnics*.

try and graphology who by then had become chairman of the Society for the Enhancement of Scientific Graphology, belonged to the second, along with Wittenberg, secretary of that association and also head of the Royal Dutch Airlines, and Spanjaard, vice-president of the Amsterdam Court. The other members of the working group combined an academic training in psychology with interest in graphology. They were Ouweleen, professor of psychology at Rotterdam, head of the psychology department of the National Railways, Wijngaarden professor and director of the Laboratory for Applied Psychology of the Free University of Amsterdam, and Van Lennep, by now professor in applied psychology at the University of Utrecht,

In this way representatives of the interpretative and the empirical-analytical approaches came together around the table to carry out a joint project. On 21 December 1955 the group held its first meeting in the Psychology Laboratory of the Free University. The minutes of that first meeting show how De Groot persuaded the other members to accept his view that carrying out research should not be the *last* task of the working group, but should in fact take *precedence* over other tasks. The other members let themselves be persuaded that the response to the ethical and training questions should be temporarily postponed and that a control study should first be made.[48]

At the very first meeting, De Groot made the empirical-analytical framework the starting point of the group's activities. He then led the group members step by step through the research scenario he had developed within that framework in recent years. First, the graphologists in the group were asked to examine, on the basis of their experience and their professional literature, what graphology claims to say with certainty about someone. These so-called pretentions would then help the group formulate hypotheses.

That the concept of research had different connotations for the other members of the group can be seen in the ways they approached this task. The subcommittee of graphologists (Wittenberg, Spanjaard, Böttcher, and Van Lennep) appeared at the next meeting of 29 March 1956 not with the required pretentions but with an extensive research plan. Graphological judgments on the basis of approximately two thousand handwritten texts would be put on hollerith cards, placed in a sorting machine, and compared to psychological assessments. They contended that "any possible affinity with certain psychological symptoms or syndromes could be shown in a purely empirical manner," as the minutes record.

De Groot dampened the group's enthusiasm: "A beginning without hypothesis is dangerous. A number of variables which seem significant can always be found. It is better to draw up one or more hypotheses whose tenability can be researched."[49] The subcommittee set to work again. The task was to isolate a single trait (variable) in order to carry out a convenient "validation" study in a relatively short period of time. Contrary to the basic principle in the *geisteswissenschaft* approach that the meaning of concepts depends on the context in which they are applied, the graphologists now allowed themselves to be seduced into selecting a single separate trait. The choice was the variable "energetic-weak." After some discussion the definitions of these concepts were chosen: "Energetic" is "someone who acts in a vigorous and firm manner and is not easily disconcerted. He shows initiative and is not quickly discouraged by disappointments." "Weak" is "someone who carries out his activities with little vitality. He allows things to run on without intervening himself. He lacks certainty in performance and his persistence and endurance are limited."

The group decided to examine in the validation study whether or not applicants for the position of advertisement salesmen, who had been judged to be distinctly energetic or weak on the basis of psychological research, were also judged to be so on the basis of a graphological interpretation. The graphologists therefore had to "predict" the psychological assessment.

At De Groot's suggestion, the execution of the experiment was assigned to his graduate student, Abraham Jansen. Ten graphologists participated in this study as test subjects. Ten psychologists who had followed a course in graphology and ten others who had not undergone such training were brought in as control groups. These thirty test subjects assessed seventy-nine people for the variable "energetic-weak" on the basis of their handwriting. The results were unfavorable for graphology: On the basis of coincidence a 50 percent rate of correct predictions or "scores" could be expected and the result was 60 percent scores. No significant difference was detected between the results of the real graphologists and those of the two control groups. Also, the correlation between the assessors (or the "reliability," as it was also called in the Netherlands then) was very low.[50]

Both the test subjects/graphologists and some working group members were critical of the experiment's design. The instruction to the graphologists was not good enough. No distinctly energetic people could be expected to apply for the position of advertising salesmen. Some manuscripts were written in pencil. There was no control group of complete "laymen" with regard to psychology. The criterion chosen was not good, and so on. Therefore the working group decided to do a new experiment. Jansen, by now graduated, stayed on to work on the new experiment. They also decided that he would complete his dissertation, with De Groot as supervisor, on the study and its methodological backgrounds.

The results of the second experiment on the handwriting of sales representatives were also unfavorable for graphology. On the basis of criticism from test subjects and working group members, a third and fourth experiment were now designed. A company assessment replaced the psychological assessment as the criterion, and an extra control group of theologians was brought in. In the fourth experiment the criticism that the study was too "laboratory-like" because it isolated a single variable was taken into account by composing a list of eighteen traits to assess. The graphologists also were allowed to keep the material in their possession for two weeks. Neither the third nor fourth experiment showed much success for graphology.

Eight years had passed. The working group members Van Lennep, Wijngaarden, and Böttcher eventually had declined to contribute further; the others had also stopped by this time. The ethical and training questions were left unanswered. Under the continued guidance of De Groot and basing his work on that of American and British methodologists like Guilford, L. Cronbach, and H. Eysenck, Jansen completed his book on the experiments and their methodological foundation. He concluded that "for judging the individual, graphology is a diagnostic method of highly questionable, and in all probability minimal practical value."[51]

Outcry over a dissertation

Both Jansen's 1963 thesis and the revised English-language version published in 1973 emphasize, as a unique aspect of the experiments, that they were set up with mutual cooperation from psychologists and graphologists. "The psychologists on the team mainly advised on methodological sound procedures; the graphologists were there to see that the experiments were fair and acceptable from the viewpoint of graphology," read the blurb on the English-language edition.[52]

However, that feeling of cooperation was not shared by the graphologists. Although they allowed themselves to be persuaded to cooperate in the working group, in light of Jansen's unambiguous interpretation of the results, they now found, on closer inspection, the course of events to be neither fair nor acceptable. From the interpretative-oriented psychologists, whom they had mainly been in contact with up to then, they had learned to attach little value to experiments in psychology. Because of the quantification and isolation of personality traits and the restriction to observable behavior, these experiments were regarded as laboratory-like and of little relevance to the practice of psychodiagnostics.

The first counterattack took place at the public defense of Jansen's thesis. At this ceremony, the professor of philosophy H. M. J. Oldewelt appeared as an advocate of graphology, and strongly opposed the thesis. He accused Jansen of having started the research with a bias, of having suppressed facts that were unacceptable to him, and of having made himself a judge rather than a reporter.

A letter by the graphologist Wittenberg appearing in the weekly *Free Netherlands* (*Vrij Nederland*) considered "the mentality of Jansen and his ilk [to be] dangerous for psychology and graphology, but above all for the interpretation which they lend to the concept of science and with which they must come in conflict with every humanistic and religious belief, and which ultimately will undermine respect for men and fellowmen."[53] The graphologists claimed in the daily papers and weeklies that Jansen had made psychology his criterion to an excessive extent, and that he had drawn incorrect conclusions from his research. They expressed their surprise that "nowadays in psychology it is possible to submit a collection of calculations to obtain the degree of doctor." They openly cast doubt on Jansen's integrity, accused him of "treating people as if they were kilos of iron," claimed that he had not yet outgrown his enfant terrible period, and that he was "sowing discord instead of knowledge." Oldewelt (whose wife had recently published a book on graphology)[54] explained to the newspaper-reading public what he consid-

Figure 6. The graphologist J. J. Wittenberg at work (photo: *De Telegraaf*, 3 March 1966).

ered to be the deficient relevance of the experiments, in the characteristic terminology of the academic holistic style:

> The graphologist allows the written material to sink in for hours, often ten hours or longer, opens himself to the uniqueness of a character. He understands the details of writing with their assorted possibilities of significance in terms of a total image which increasingly loses its vagueness without becoming a mosaic of impersonal qualities. And when he finally writes down his findings, the task of his words is to add nuances to each other in our lives while remaining a whole. A neutral definition is of no use to them. Their power and clarity are in the report as a whole, not in the lone contribution of each. This is not something mysterious: The sounds which go to form an indivisible melody do the same. Every sen-

tence which we speak to each other, despite the tendency to separate meaning of each individual word, is nevertheless a whole,
attuned to the current situation and therefore having a unique meaning in that context.[55]

Jansen himself responded to his accusers in the various newspapers.
The Dutch Association of Practicing Psychologists also entered the fray
on his behalf. It sent a notice to the press that "Dr. Jansen's study was
carried out on objective scientific principles" and that "such studies,
which are also carried out on other diagnostic instruments are, irrespective of the results, of importance to the development of the field concerned and of science." The association, which some years earlier was
making plans to promote graphologists to psychological assistants, now
took up the cause of Jansen's defense and did nothing to support the
graphologists.

In general, the psychologists who had formerly consorted with graphology now began, though somewhat hesitantly, to distance themselves
from it. For example, working group chairman Ouweleen, in his final
report to the Society for Labor Psychology, wrote that "with this question, with this way of working and with the various criteria chosen for
this study, it has not been possible to show that graphological statements
of each participant show a significant correlation to the chosen criteria."[56] In Van Lennep's Foundation, according to an internal report, "the
self-image of an institute which is ahead of its time, has been dented" by
the graphology affair.[57] Ex-working group member Wijngaarden also began to have his doubts: Graphology was given a far less prominent place
in the Amsterdam Laboratory for Applied Psychology.[58]

The psychologists who had still supported graphology shortly before
were now so uncertain that they no longer openly came to the assistance
of their former protégés. Therefore public support from the academic
world was confined to that from the philosopher Oldewelt. By distancing
themselves from interpretative thinking, the psychologists, unlike the
graphologists, did not lose at one stroke their entire future in psychodiagnostics. They could afford a quieter changeover.

In the 1950s only a few articles and books in the empirical-analytical
style would appear,[59] but by the 1960s the profession had moved closer
to completing the "gestalt switch" to empirical-analytical thinking. In
1961 De Groot published his book *Methodologie*[60] and at almost all
Dutch universities psychologists began to make a case for an empirical-
analytical way of working, both with regard to diagnostics and with
regard to methodology in the broader sense.[61] From Nijmegen University, the first students left for the United States to acquire knowledge of

statistics and measurement theory. The Americans E. L. Kelley and D. W. Fiske, L. J. Cronbach, G. C. Gleser, J. C. Nunnally, J. S. Wiggins, and P. E. Meehl became the big names in Dutch psychodiagnostics.

In 1969 the Test Research Committee (*Test Research Commissie,* hereafter TRC), set up by the Dutch Association of Psychologists, published a report rating the various tests in use. In these test evaluations the requirements of validity and reliability were the most important norm. The marking of the tests ran from A as highest score to F as lowest. Interpretative tests like Van Lennep's Four Pictures Test and the Rorschach did not receive anything higher than a D in this report (and only two of *all* the tests then in use got an A, whereas half of them were not even included in the rating system but ended up on a list of rejected tests at the back of the report).[62]

Nowadays the unscientific nature of *geisteswissenschaft* thinking in Dutch selection psychology is almost undisputed. That the concept of science refers to an *empirical-analytical* definition of science is now so much taken for granted that it does not even need to be defended. This is also the case in the way people think about graphology. Whereas, for example, until 1964 the well-known *Textbook of Psychology* (*Leerboek der Psychologie*) stated that it "was above all L. Klages who managed to raise graphology to a science,"[63] from the revised fourth edition of 1968[64] onward we are told that research like that of Jansen's leads to the conclusion that graphology cannot be considered as a scientific method. In 1980 the well-known contemporary psychologist-methodologist Hofstee (see Chapter 5) praised Jansen's study again as having been carried out "in the spirit of Popper" because of "the serious attempts" that Jansen, despite his skepticism about the value of graphology, "made in collaboration with the graphologists to prove the predictive value of graphological statements."[65] And an introduction to psychology from 1985 states briefly: "There are graphologists who are skeptical about an experimental testing of their methods. With their arguments, which we will not go into here, they place themselves outside scientific discussion."[66]

The social identity of the assessment psychologist

No matter how unsound the basic principles and working methods of interpretative psychology may seem from a contemporary viewpoint, this almost general switchover to empirical-analytical thinking cannot be simply explained by arguing that Dutch psychologists only gradually came to understand how their profession had to be practiced in a scientific manner. Such an explanation would ignore the fact that, if not graphology on its own, then *geisteswissenschaft* testing on the whole was once *broadly* accepted. Van Lennep's Foundation for Psychotechnics alone included 1,200 companies among its clientele in 1946, as well as 40 government bodies and 25 schools. At that time the Foundation was employing forty-four people.[67] This success was achieved although alternatives closer to our contemporary methods were offered by bureaus like that of Luning Prak. To see the changes as a slowly evolving insight into the correct approach would imply that both the early interpretative psychodiagnosticians themselves and their clientele were quite simply suffering from mental blindness. Since that is not very plausible, another answer must be found to the question of how it was possible that interpretative methodological norms set the agenda from the 1920s to the 1960s. After that, an answer must be found to the question of how in the following decades empirical-analytical rules became the standard for a rational assessment psychology.

How could the interpretative psychologists defend their empathy-based approach with so much self-confidence and success until late in the 1950s? How could they convince themselves and others that the scientific nature of their statements was guaranteed by good intuition, much experience, and a high level of training? What was behind their idea that experimental research was too laboratory-bound and in conflict with "humanitarian principles of living"? As I argued in Chapter 1, the study of the process in which a (sub-) discipline sheds an old scientific identity and acquires a new one requires the investigation of the altered *social identity* of this (sub-) discipline. It was assumed that to achieve this end it is necessary to examine who were the most important clientele of the profession and what were the presumptions current among those customers.

To accomplish this goal, we must take a step back in history. My story of the early twentieth-century psychotechnicians and their clients must begin in the second half of the nineteenth century. At that time, industrialization, which started later in the Netherlands than in other European countries, was under way. Historians of labor have often noted that just as much poverty and wretchedness arose among the Dutch proletariat as in England, France, Germany, and Belgium, but in the Netherlands no revolutionary class was founded.[68] When trade unions began to appear in the last decades of the twentieth century, radical socialism failed to take hold. Socialists in the Netherlands mostly sought to transform the means of production from private to public hands through the peaceful humanization and democratization of capitalism, rather than through revolutions. Furthermore, in the Netherlands the denominational trade unions, which were based on the basic principle of *harmonious cooperation* with employers, attracted far more members than in the neighboring countries. The Catholic workers in the south of the Netherlands joined Catholic unions and the Calvinist workers in the north joined the Protestant unions.[69]

Moreover, Dutch society in general came to be segmented or, as sociologists came to call it, "pillarized." This concept refers to the phenomenon that life for the great masses took place in all respects within the specific social subgroup to which each individual belonged. Not only did socialist, Catholic, Protestant, and neutral trade unions emerge, but socialist, Catholic, Protestant, and neutral political parties, youth organizations, schools and universities, welfare institutions, gymnastic associations, newspapers, and (later on) radio and television associations were established as well. In this way people hardly came into contact with people of other denominations. Also, the electoral system was designed according to the principle of proportional representation. Whereas the common man voted for the representative of his own particular group and in other respects humbly and respectfully did what his leaders laid down for him, these representatives worked together to arrange the *general* social matters of the nation in mutual consultation. This led to a system of political autonomy within the various groups, cooperation between the leaders of these groups, and submissiveness and passivity among the broad masses. As a consequence, there was a high degree of political stability. In the English-language literature, this system is called "pacification democracy"[70] or "consociational democracy."[71]

Early assessment psychology emerged in this social climate. Unlike the situation in other countries, many of the first Dutch psychotechnicians were not employed in education, business, or the army, but rather by one of the many institutions that applied themselves to the amelioration of the social consequences of industrialization.[72] An early example was the psychotechnical bureau of the Roman Catholic labor organizations directed by the Jesuit Van Ginneken (founded in 1919). Immediately after this institute was founded, the neutral Dr. D. Bos Trust (already mentioned) was established for the "undertaking or supporting of sociopedagogical work in the broadest sense of the word." The psychologists Brugmans and Prak came to work here. In 1921 the Protestant Christian Psychological Center for School and Vocation (*Christelijke Psychologische Centrale voor School en Beroep*), which from 1928 was run by Jan Waterink, a theologian and professor of psychology at the Calvinist *Free University* was founded. In 1921 the municipal bureau in Amsterdam had also followed, as the Utrecht municipal bureau (1924), whose director was the theologian and later psychologist Van Lennep. In 1927 this bureau became the Dutch Foundation for Psychotechnics.

In these welfare institutions, organized psychological support was considered necessary, now that the differentiation of functions and the demands of urban life had deprived workers of their vocational pride, and had destroyed the foundation of the unwritten moral rules that had regulated life in the small communities that were now in decline. However, although these institutions were inspired by indignation about social abuses, they did not have any *revolutionary* zeal. They thought in terms of harmonious relations between labor and capital and not in terms of clashes of interests. Their ideas fitted into the general sociopolitical climate marked by the attempt to neutralize the negative aspects of industrialization, mainly by promoting adaptation to the new structures.

Moreover, because labor and capital were supposed to have shared interests, it was taken for granted that they could be served concurrently and with the same means. Testing applicants for a company was seen as a form of help for the job seeker. Working in a welfare institution therefore was felt to be completely harmonious with offering service to industry. As Brugmans expressed it: "The requirement of the right man in the right place can be motivated equally by principles of an economic or humanitarian nature. This requirement is good, also in the moral sense of the word. Do not many people lead a sorry existence because they are not up to their work?"[73] Or as Roels put it: "Employers and employees will benefit equally from it: The workers who will occupy positions where their physical and mental capacities are done most justice to, the employers who acquire workers who are suited to their task."[74] And Van Det, the leader of the Amsterdam municipal bureau, summed it up as the "economic-pedagogical task" of his institute.[75]

All of this created the situation that a Dutch psychotechnician could simultaneously be an applicant's assessor *and* his personal advisor. This also explains why an assessment psychologist was looked upon most often as a *spiritual mentor* rather than as a detached observer. The psychotechnician was watched over by "the unimpeded influence of religion on the investigations"[76] and used psychotechnical means "which God has bestowed upon us in this time"[77] to help one's "fellowman" find a place in earthly existence.

This social identity of the psychotechnician as a spiritual mentor was even strengthened when industry began to ask for the presence of special *character traits*. Because of the growing division between ownership, management, and clerical work, the group of white-collar workers was growing. Now, the "right man" could no longer be defined only in terms of his sensorimotor skills. For management, the secretarial office, the sales department, the international correspondence office, and the drawing section, a demand emerged for personnel with particular *psychological* qualities as well. Attention had to be paid to positive or negative character traits such as "ambition," "trustworthiness," "perseverance," "vanity," or "punctuality."

American psychologists had designed questionnaires for the assessment of character traits. The validity of those questionnaires was determined by calculating correlations between patterns of answers and traits defined in terms of overt behavior. This approach was known in the Netherlands but it was rejected on ethical grounds. The pedagogical atmosphere in which psychotechnics was embedded ensured that the required psychological traits were regarded as *moral qualities*. In the Christian bureaus especially it was emphasized that an individual could not be regarded as a "thing with traits," but that he or she was a "person" (with a soul).

For example, on the occasion of the tenth anniversary of the Amsterdam bureau, Roels, a Catholic, argued for a new psychotechnics that would no longer "dissect human beings into pieces." He connected that idea as follows to the elevated task of personnel assessment:

> because the practical psychologist is not someone who . . . has to do a number of experiments on an individual in order to then see if the list of traits is correct. No, the vocational advisor who does his job properly is a pedagogue! In the first place he tries to adapt what he considers necessary for a particular individual to the special nature of the individual. If the vocational advisor does that he will be more than a pure analyst; he is then a synthesist. However, he cannot be the latter without the addition of love, which he must have for the individuals he is dealing with and about which Paul writes so beautifully in the letter to the Corinthians, saying that love is what believes all, hopes all and endures all; that love which will still burn when all the prophecies and sciences have come to dust. That is the synthetic task of the advisor on a choice of career![78]

It was considered necessary to lay bare all "inner aspects" of a job applicant. Van Lennep, whose Foundation had specialized in assessments for higher positions, contended that "the last, the deepest, the most essential aspect of a person's being in the world" should be investigated. The question of what could be "higher than intuitive judgement of each other" was posed and psychologists spoke of "pervading the person."[79]

It was this pedagogic approach to the task, this professional image of the psychotechnician as a mentor, that was expressed in the sensitive, intuitive approach and in the definition of the psychologist as a person of wisdom and experience. The "self-disclosing" methods of interpretative psychology helped applicants to reveal themselves and helped the psychologist to gain a "deeper insight" into their personalities than they could achieve on their own. The validity or predictive capacity of test instruments was irrelevant in this context. Besides, as Van Lennep stressed, "schematizing and observing by fixed rules" was "harmful to intuition."[80]

This provides a two-step answer to the question of how it was possible that the interpretative approach could establish itself. First, it was not the rise of a particular method for a constant and precisely defined question that was involved in the founding of this approach, but a *specific interpretation* or a *transformation* of that question. In continuation of the early psychological laboratory research, the focus of assessment psychology was originally on cognitive, motor, and sensory capacities, but it was soon argued that in fact the (moral) essence of the individual must be exposed. Next, the question of how these new views could take hold was answered by pointing out that this specific interpretation of the selection problem was rooted in the labor relations that existed in the Netherlands of that time. Through the general belief in harmonious relationships between workers and management – a belief that was part and parcel of "pillarized" Dutch society with its pacification democracy – the aims of the business sector could be served in the welfare institutions (where the early psychotechnicians were working) and the aims of the welfare institutions were thought to be served by industry. This is how the interpretative methodological style with its empathizing approach could make its way in psychological personnel selection. It adequately reflected and gave shape to the dominant ideas about correct industrial relations.[81]

The next question, provoked by this depiction of events, is how was it then possible that empirical-analytical rules became normative in the 1960s and 70s. The answer to this question requires some background on the social history of the Netherlands in the first decade after the Second World War.

Changing methodology in a changing society

In the early postwar years Dutch socialists, who had taken an even more moderate course than they did in the prewar years, acquired governmental power for the first time. However, the country was in a bad way. Roads, dikes, mines, and factories had been bombed or plundered. Productive capacity had been reduced by about 40 percent. The government decided upon "centralized wage determination," in which the level of salaries was not determined by negotiations between employers and employees, but kept under central control. Despite the efforts of a number of people to break up the "pillarization" of Dutch social life (more on this in Chapter 3), the segmented society remained as it was before the war. As before, voting behavior was determined by the group a voter belonged to and for the rest matters were settled by the elites. Such factors meant that in this period, too, no labor movement arose that opposed employers rather than collaborating with them.

Nevertheless, some things were gradually changing. For instance, in 1950 the Parliament, inspired by the American "Human Relations Movement," adopted the "Works Council Act." Although the notion that employees would get a real voice in company policy met with such resistance from employers that this possibility was not guaranteed by the law, it was nevertheless stipulated that every company with more than twenty-five employees had to set up a "Works Council." In 1953 productivity once again reached the prewar level and in the following years employment recovered and wages grew. At the same time, employees began to protest against centralized wage determination and demand more participation in company policy. The educational level of the population also gradually grew and the means of communication increased.

In the 1960s, changes occurred more rapidly. The economy advanced further. In some periods there were labor shortages. Complex processes of change were under way, but those processes can only be broadly sketched here. The Dutch themselves use concepts like "depillarization," "deconfessionalization," "polarization," and "politicization" to describe these societal changes. Ideological bonds and with them traditional political bonds grew looser. People became more emancipated; they no longer did, true to tradition, what their leaders said, and they began to participate more in political affairs. As the Czech sociologist Ernest Zahn has noted, there is an irony in the fact that the "pillars began to totter" in the Netherlands precisely in the period when the *image* of accommodation began to penetrate into the Dutch political consciousness.[82] Indeed, the Dutch political scientist Arend Lijphart, who had written his book *The Politics of Accommodation: Pluralism and Democracy in The Netherlands* in the mid-1960s in California, already felt compelled in 1975 to add a chapter to the second edition entitled, "The Breakdown of the Politics of Accommodation." In the foreword to the second edition Lijphart wrote: "It is now clear that the developments of the mid-1960s were merely the beginning of rapid and far-reaching changes that challenged the very foundations of the politics of accommodation. In fact, with the wisdom of hindsight, I now think that the politics of accommodation in the Netherlands came to an end around 1967."[83]

Things were also changing in the sphere of industrial relations. The "workers' control" sought by some was never achieved, but nevertheless strikes and conflicts increased and in general harmony decreased.[84] In 1968 it was legally laid down that employers and employees again had the right to determine working conditions through mutual negotiation. Although various writers on these developments warn that the changes should not be exaggerated and although at the end of the 1970s "restorative tendencies" were again noted, it is generally agreed that there was a shift in the relations of power. Gradually, the climate of organized consultation *(overleg)* was exchanged for that of negotiation *(onderhandeling)*.[85]

These developments also found expression in the prevailing ideas on the allocation of people for jobs. With the rising level of education, more people began to consider themselves capable of making personnel assessments, someone's social status was no longer sufficient guarantee for the value of his ideas, and the empirical validity of a judgment began to be more important than the pedagogic content. "Five Times Tested, Five Different Recommendations: In the Expensive Maze of Psychotechnics" was the headline above a 1950 article in a popular weekly. The author gave an irreverent account of his experiences at five psychotechnical bureaus and of the resulting reports.[86]

More such incidents were to follow in the 1950s. Selection research could no longer be seen as an aid "bestowed by God in our time" on all, but instead gradually came to be looked upon as an instrument that served only the employers' interests. In 1955 a socialist daily published a series on the practice of selection entitled, "Job Application: On Show, Unprotected."[87] And a headline above one of the articles stated plainly: "Our People Are Being Exposed to Mental Charlatanism." The paper denounced the fact that the question "what can you do" really was "who are you," that the reports were sent to the employers, and that for years they were taken out of the cabinet by the companies when decisions had to be made again.

In the course of the 1960s, critical voices on behalf of job applicants grew in strength. In 1966 a large national newspaper devoted a special issue to personnel selection, which spoke of "the smiling recruiting officers of the boom."[88] Whereas in 1952 the "Welfare Plan" of the Dutch League of Trade Unions had still mentioned the "great economic and social importance" and the "essential contribution to the increase of job satisfaction" of psychotechnics,[89] a different tune was heard in 1970. "To hell with psychotechnics," stated a staff member of the recently established "Scientific Research Foundation of the Trade Unions."[90] "The practice of testing is one of the many tokens of an industrial relations system, marked by servitude," he wrote in a national daily under the heading of "Feudal Test."[91]

Nevertheless, matters did not come to a head in an "antitest movement" such as what began in the late 1950s in the United States.[92] In the Netherlands the new generation of university-trained psychologists was able to translate the criticism successfully into protests against only the *geisteswissenschaft*-oriented practice. They believed and argued that it was the basic principles of this practice that were the cause of the problems, and they presented *themselves* as spokesmen of the resistance. The effort to expose "the most essential of someone's being in the world," they argued, is an unacceptable attempt to extract intimate information from a job applicant, in the interests of someone else. For example, in a weekly, a professor of labor psychology called on job applicants to act in a more assertive manner, and instead of putting on their "Sunday suit" for an application, they should protest against the "excavation of their personality."[93] The new generation of empirical-analytical psychologists also criticized the interpretative working methods used by the National Psychological Service (*Rijks Psychologische Dienst*) in the selection of civil servants. A professor in selection psychology spoke in an interview of "pure witchcraft" in relation to the methods of this Service.[94] The Dutch Association of Psychologists also lectured the Service: The way they handled applicants and reports was in conflict with the ethical code of the professional association.

At the same time, however, these psychologists argued that *good* professional psychodiagnostics could actually avoid handing over applicants to the subjective interpretation of higher-ups. It was professional psychodiagnostics that could safeguard applicants against in-depth research into their persons. A condition for this would be, first, that not everyone is allowed to carry out selection research: Only those who use validated research instruments with which purely practical information can be obtained, and who above all are subject to a professional code with ethical rules could participate. If an assessor meets these requirements, it was argued, the applicant is better off with psychological testing than with the randomness and the preconceptions of personnel heads and others who are not experts in the field of selection.

So, just like the psychotechnicians of the 1920s and 30s, the assessment psychologists of the 1960s and 70s offered their professional service not only to the business sector, but also to the employees. It is intriguing that in the years between the wars humane treatment of the applicant was the argument used to *introduce* interpretative methods, whereas in the postwar decades "the rights of the applicant" was the argument for *expunging* these methods. In the 1920s social circumstances led psychologists to turn against their "elementaristic" colleagues, who according to the criteria of that time, did not do full justice to the subjects, but in the 1960s progressive psychologists turned against their interpretative-thinking colleagues, who according to the *new* social criteria, were causing damage to the interests of the applicant. In a 1976 book that was even dedicated "To the Vulnerable Applicant," psychologists gave an account of their struggle against colleagues for a just assessment psychology.[95]

The arguments of these psychologists turned out to be convincing. For example, in 1973 a motion in Parliament on psychodiagnosticians' unnecessary deep intrusion into the private sphere led to the establishment of a committee that studied the desirability "of measures to protect the personal interests of employees during the procedures of recruitment, selection, and appointment." The former graphology researcher, Abraham Jansen, who by now had become director of the Amsterdam para-university Institute for Social and Industrial Psychology, was appointed to this committee. In its 1977 report (entitled *An Applicant Is Human too*), the committee emphasized the right, among other things, to an "instrumental, purposeful procedure." That meant that selection instruments had to meet the methodological requirements of validity and reliability. In this fashion the selection procedure, according to the committee, would contribute not only to giving applicants a fair chance, but also to protecting their privacy. In the report, graphological research is also explicitly rejected as an autonomous selection instrument, remarkably enough not because of its imperfect validity but "because privacy is disturbed in an unacceptable manner." The report says psychological research "can make a valuable contribution in the selection of applicants."[96]

It thus appeared that with the transition to the modern empirical-analytical methodology as was the case before with the transition to *geisteswissenschaft* methodology, the methodological upheaval simultaneously involved a transformation or new specification of the question of suitability for a job. This was no longer a question of what someone essentially *is*, but what he or she *can do*. And also, the new interpretation now rendered expression and form to specific ideas on correct labor relations. When divergent interests were ascribed to capital and labor, applicant testing was no longer seen as a help to the applicants, employers were deprived of the right to know "everything" about them, and other methods came to set the tone in selection psychology.[97]

This story could also be summarized by saying that the earlier interpretative psychologists lived in a *world different* from that of their present-day colleagues, not merely in the methodological sense but also in social terms. Instead of the model of the social worker, they presented that of the businesslike, detached middleman. The assessment psychologist became a broker in manpower paid by the buyer, who claims simultaneously to serve the interests of the seller. In this fashion the profession was able to establish itself, right up to this day, even though – as the metaphor of the broker working primarily for the buying party may make clear – some protest from the employees' side is registered from time to time.

So this history contradicts the simple picture in which the interpretative psychotechnicians did not yet properly understand how "the" profession of the psychodiagnostician should be "scientifically" practiced, whereas the postwar psychologists gradually came to understand this better. The changes in the methodology of selection psychology were not the consequence of growing insight into independent criteria for a correct approach, but were the result of the fact that the profession itself changed as a component of a changing society. The change of *scientific identity* pointed to a change of *social identity*. Therefore it is not unthinkable that, together with possible new changes in social relationships, the present-day methodological criteria of assessment psychology will also someday be subject to revision.

3

"Like everything living which encounters us"

In 1973, the historians of psychology Henryk Misiak and Virginia S. Sexton noted that "a large proportion, if not most, of the influence of European phenomenology on American psychology in the last two decades has come via Holland."[1] Apart from the work of Misiak and Sexton, Dutch phenomenological psychology is also the subject of much interest in other English-language surveys.[2] Indeed, there was a time, roughly from 1946 to 1960, when a phenomenological psychology was developed in the Netherlands that managed to attain a strong national and international position. That took place mainly at the University of Utrecht, where Frederick J. J. Buytendijk was the prime mover of the new approach.[3]

Buytendijk was appointed professor of psychology in 1946, but that appointment was made by no means as a matter of course. Right up to the present, Dutch historians of psychology recall the widespread amazement this appointment created.[4] The fifty-nine-year-old Buytendijk was not a psychologist but a physiologist and physician. His only psychological works until then had dealt with animal psychology; he had written on topics such as "The wisdom of the ants," "The mind of the dog," and "methods for psychological research of invertebrates."[5] Openly and "quasi-apologetically" he admitted never to having looked into a manual on human psychology: "I am more than anyone a stranger in the field of psychology."[6]

The professorial chair became vacant when Buytendijk's predecessor, F. J. M. A. Roels, was removed immediately after the war for collaborating with the Germans. It was the educationalist and psychologist Martinus J. Langeveld who then got Utrecht psychology back on its feet again. He in particular judged Buytendijk to be a suitable candidate for the vacant chair and he brought about Buytendijk's appointment with the help of the influential Nijmegen professor of psychology Franciscus J. T. Rutten,[7] who in the same year arranged for Buytendijk to become professor of psychology at his own university.

I mentioned in the previous chapter that Van Lennep was appointed professor of psychology at the University of Utrecht in 1949. Although the literature reveals little surprise at this fact, that appointment was actually just as unlikely. It is true that Van Lennep, a theologian, had also graduated in psychology in 1942, but he too had little affinity for other psychologists. As I already related, as a head of a private institute for personnel selection and vocational guidance, the Foundation for Psychotechnics at Utrecht, Van Lennep had even refused to appoint academically trained psychologists. To his mind academic psychology destroyed rather than enhanced people's ability to judge other human beings. Not only did a number of psychologists attached to the university consider Van Lennep's methods inferior, but above all they disapproved of his appointment policy in the commercially very successful Foundation. Thus in 1939, when Van Lennep had applied for membership of the NIPP (the Dutch Association for Practicing Psychologists), membership was denied him, despite the fact that at that time it still was possible to become a member if one had not graduated in psychology, but had just been working in the field of applied psychology.[8]

Archival material also reveals that the Utrecht faculty board was quite unenthusiastic about Van Lennep's appointment. They strongly recommended the appointment of the Amsterdam psychologist De Groot. Van Lennep had no Ph.D., it was argued, and he had published hardly anything, whereas the much younger De Groot already had a doctorate and a substantial list of publications.[9] However, Langeveld and Buytendijk definitely wanted Van Lennep and not De Groot. Therefore Van Lennep had to get his doctorate like "greased lightning," as Langeveld related in 1980.[10] He took a study leave from his activities at the Foundation and finished his dissertation in 1948, under the psychiatrist H. C. Rümke, on the phenomenon of projection and projection tests.[11] And with the support of Rutten, who in the meantime had become Minister of Education, Art, and Sciences, Van Lennep was eventually appointed two years after the procedure began.

As shown by this appointment policy alone, in the postwar reconstruction of Utrecht psychology a conscious attempt was being made to do something completely different than what was called "psychology" elsewhere. The "natural scientific" or "positivistic" approach was rejected in favor of so-called phenomenological psychology. At the beginning of the 1950s the conceptual framework of this Utrecht phenomenological psychology came quickly into vogue among some Dutch psychologists, who spoke and wrote of "the experienced," "the encounter," and "the I-Thou relationship." Books appeared with titles like *Person and World (Persoon en Wereld)*,[12] *Rencontre, Encounter, Begegnung*,[13] and *How Do I Encounter My Fellowman? (Hoe Ontmoet Ik Mijn Medemens?)*[14] In Utrecht at this time the phenomenological approach was emphasized so much that soon there was talk of "the Utrecht School."

The concept "Utrecht School" is not clearly defined by a particular group of people. To begin with, the designation was already in use with regard to a circle of Utrecht phenomenological *criminologists*, led by W. P. J. Pompe. Although there certainly were links between psychologists and criminologists, they were nevertheless two distinct groups. A number of psychiatrists such as H. C. Rümke and J. H. van den Berg could also be mentioned in connection with the Utrecht School, and the sociologist J. P. Kruijt. The phenomenological psychologists' links with the Utrecht psychiatrists were closer than their ties with the criminologists. Some of the psychiatrists gave a subsidiary course for psychology students and together with the psychologists they published in collections. However, in this chapter the Utrecht School mainly refers to the central group in *psychology*. This group consisted of Langeveld, Buytendijk, Van Lennep, and – from the younger generation trained there – Johannes J. Linschoten,[15] and somewhat more tangentially, Benjamin J. Kouwer and Jos H. Dijkhuis.

The central object of research here is only the phenomenological thought of the Utrecht *psychologists*, and more specifically the methodological aspect of it. This phenomenological methodology will neither be attacked nor defended. There has already been extensive debate in psychological circles for and against phenomenology.[16] And partial historical works on this topic have also been written,[17] whereas, as far as I know, up to now none of the many phenomenological trends and schools in the history of psychology has been analyzed on the basis of a detached and contextually oriented approach. This in itself seems reason enough to lend impetus to it. As I explained in Chapter 1, though, it is mainly because of the *goal* of my research that I have decided to adopt the position of a disinterested outsider with regard to methodological correctness. The aim of this chapter is again to illustrate empirically the thesis that subscribing to particular professional rules involves choosing specific positions on the level of social relations. I will show, this time on the basis of Utrecht phenomenological psychology, that the degree to which a methodological style is accepted and maintained varies with the contemporary adequacy of the social identity expressed in it.

For these reasons, I also do not pretend to present "*the* history of the Utrecht School" in this chapter. I do not offer a comprehensive view of the school, let alone describe the contributions of each separate member. My aspirations are more modest. First, the methodological aspects of Utrecht phenomenological psychology are worked out so that I can then investigate how these psychologists came to hold their specific viewpoints on psychology. This in turn clarifies how it was possible that the phenomenological psychology of the Utrecht School could flourish precisely in the first postwar period in the Netherlands, whereas in the 1960s it was quickly forced back by the statistical empirical-analytical approach.

The methodological basis of the Utrecht identity

Even though the psychologists of the Utrecht University did not speak themselves in terms of a "school" and even though their working methods were often divergent, they did give their psychology its own name and they also had a strong *awareness* of their own identity. Both the teachers and the students explicitly distanced themselves from other, nonphenomenological psychologists, who were considered brutal and superficial. For example, Buytendijk contrasted "natural scientific psychology" with "phenomenological psychology" as follows: "The living creature, owned by no one but himself, does not need to become the prey of the psychologist that he tears apart and digests or conserves and pins down in solid concepts. The inner, like everything living which encounters us, can also become beloved, be looked at and cared for tenderly."[18]

In letters the professors wrote to each other, they gave each other compliments on articles or lectures, in which they encouraged each other to continue the shared resistance to the positivist deterioration of psychology, and they quoted from philosophy and world literature.[19] They also had a monthly meeting, called the "cercle" (in French) or "the phenomenological discussion circle," to which students were not admitted. The development of a phenomenological identity among the students, however, was promoted rather than impeded by the distance the teachers kept from them. In many lectures students were told that one does not just "become" a phenomenological psychologist, but that it required a process of maturing; the students were the chosen few who would belong to a select group bearing great responsibilities. Sometimes Buytendijk addressed a student, to the student's great pride, as "my son." The students had a society paper called *The Phenomenon (Het Verschijnsel)* and at the annual national conference of psychology students they sang: "In Utrecht we study psychology, and see the true

Figure 7. F. J. J. Buytendijk in 1960 (photo: F. J. Dehue).

essence of phenomenology. The others are just chattering and do not know anything. It is evident to each student that it cannot go on like this. Phenomenology will be all that is."[20]

Agreement as to what psychology should *not* be gave the Utrecht psychologists a strong basis for identity. The natural scientific psychology, represented by the University of Amsterdam in particular, served as a negative frame of reference. The methodology of De Groot, who after missing his chance in Utrecht became professor of psychology in Amsterdam, was sharply condemned by Langeveld, for example. Langeveld applied the following qualifications to it: "exclusive and dogmatic," "the most irrational outrageous demands," "a tragic misunderstanding," "nomological dictatorial system," "scientific stranglehold," "inner weakness," "untenable proposition," "discreditable dogmatism," "exaggerated pretentions," "a-psychological," "anti-creative," "paralyzing," and "completely unacceptable." And all that was on one single page.[21]

An obvious question therefore is, what methodological *alternatives* did phenomenological psychology offer? However, after looking at earlier authors who had sought clear and coherent rules for carrying out research among the Utrecht phenomenological psychologists, this question seemed easier asked than answered. In an obituary of Buytendijk, who quickly had become the real leader of Dutch phenomenological psychology, his contemporary and colleague Rutten had this to say:

> How did he carry out that research, how did he track down the characteristics? I asked him about it. He came back to the question three times. In so far as I understood the answer, it boiled down to this. A phenomenon must be examined in its dependence and interdependence on the concrete circumstances in which it occurs, from various viewpoints, cultural-historical, genetic, physiological, and in relation to phenomena which are found together with it and are related, also in its pathological forms. I was not much enlightened about his observation technique either. The only thing he said on this was: "by holding a phenomenon up to the light its hidden structure is revealed." He had a patent on his method. His basic stance was observational thought. . . . The way in which he absorbed the observed remains the Buytendijk secret.[22]

A historical study of the Utrecht School, made in the 1980s, even concludes: "The Utrecht School never made it clear what was phenomenological research and what was not. . . . There was nothing like a phenomenological research program. 'First look and consider well,' could be a brief description of the method of the Utrecht School, but that was then developed differently by everyone."[23] And a third author simply claimed that what went on in the Utrecht School was nothing more than "the exchange of (prewar) parochial language for obscurantist jargon."[24]

On reading the texts of the Utrecht group I also could not figure out at first what *else,* in methodological terms, the members had to offer besides their constant harping about the objectionable character of "natural scientific psychology." Initially there seemed no other identity to discover than this negative one. *If* methodological concepts were used, they were applied in ways that now appear strange. For instance, in contemporary methodology "objectivity" stands for taking measures for weeding out the researcher's personal preconceptions, but in the Utrecht School the "natural scientific" psychology was called "objectivist" and the real objectivity of the phenomenological psychology was then defended: "Because it does justice to man's subjectivity, it considers itself an objective science: true, in its scientific research, to the nature of its 'object.' "[25] Methodological indications for the way in which "fidelity to the nature of the object" should be accomplished, however, were not to be found.

It seemed as if the existence of a school in the past of Utrecht psychol-

ogy was nothing more than an artifact of historians, summoned into being
by a mere proclamation of a new psychology. My doubts on this matter
continued to exist for a considerable period of time, until suddenly I
realized that despite my intentions not to impose contemporary norms on
the past, I was still reading the school's texts too much in terms of the
framework of contemporary empirical-analytical psychology. The search
for alternatives to the *indications* of the contemporary empirical-analytical methodology was, I realized, not the way to shed my contemporary
lenses; rather, it was precisely because of this search, that methodology
was still functioning as the frame of reference. In *empirical-analytical*
thought, detailed rules for empirical research are an essential characteristic of science. Here, there are rules for formulating hypotheses, for
setting up experiments, for calculating results, for drawing conclusions,
and so on. In the phenomenological psychology of Utrecht this was not
the case. The Utrecht psychologists agreed on matters of a completely
different *nature* than those of contemporary psychology. Anyone who
wonders, with respect to their approach, what the alternatives to the
present-day standards are and looks for other rules for setting up and
carrying out research, will therefore find little coherence in their work. A
gestalt switch is required in order to understand the school's program.[26]

According to the Utrecht School, one did not become a phenomenological psychologist with the help of manuals on methods and techniques,
but rather by becoming a more mature type of person, with a natural
insight into and an unforced choice for the good. In this way one acquired
the capacity to really understand. That capacity does not express itself in
the gathering of data but is characterized by a loving participation in
what exists and, in the case of psychology, in the existence of other
human beings. In this context, the other arrives at self-knowledge and
full humanity through dialogue with the psychologist. Liberation from
fears and feelings of guilt should encourage people no longer to transgress
morality out of incompetence, or keep rules out of blind obedience, but
to choose freely that which is the purpose of life.

From this perspective, it is for instance no longer strange that the
concept of objectivity is connected to fidelity to the nature of the object.
In this mode of thought the methods of knowledge are not based on
epistemological principles but on ontological and ethical ones. The doctrine of being precedes the doctrine of knowledge. Insight into the anthropological and ethical conditions of being human *implies* insight into the
way in which one can arrive at psychological knowledge of people, and
is superior to it. In Utrecht thought, to put it another way,
an epistemological statement was *eo ipso* a moral-anthropological
statement.

This is also illustrated by the quotation from Buytendijk given earlier in this chapter. The natural scientific psychology was mainly rejected in ethical terms: It "digests" and "tears apart" and does not treat people in a loving way. Buytendijk made this judgment in his aforementioned inaugural lecture at Utrecht in 1947.[27] To know the innerness of the other, on this occasion he also said, arguing ontologically, in "real psychology" there are two methodic principles that are based on two *modes of being* of man. Both are distinct from the natural scientific approach to phenomena in which a human being does not judge a human being, but reason judges matter. The first methodic principle is the observation of human activity as something that has meaning. This is based on the mode of being of the "attentive existence in the world," the existence in everyday life. However, someone's "innerness" cannot be known on this basis. According to Buytendijk innerness can be known only through the second mode of existence: "Human existence does not proceed as merely a being thrown into the temporal world of cares, but is also . . . a being sheltered in the innerness of the heart, which is not of this world." This second mode of being is the basis of the method that Buytendijk called *the encounter*. This method is characterized, as Buytendijk expressed it, by the "disinterested and desire-free yet personally interested participation in each other." Through the encounter a knowledge is achieved "which absorbs the purely discursive and intuitive and transcends them."

Two years later, when it was Van Lennep's turn to give an inaugural lecture, he was inspired by Buytendijk's. Van Lennep also related ways of knowing to anthropology and ethics.[28] He described "three methodological viewpoints," not in epistemological terms as is customary in the empirical-analytical tradition, but as three types of relations between the psychologist and the subject studied. What concerned him was "what, in the various methodological standpoints which the psychologist can adapt, is his relation to an other, individual person . . ."[29] In doing so he distanced himself from the methods of the *geisteswissenschaft* tradition to which he formerly adhered, and stated his preference for the new phenomenology. The method of understanding from the *geisteswissenschaft* tradition in fact makes the "fellowman" hardly less of an object than the measuring methods of the empirical-analytical tradition, Van Lennep argued. Only the encounter provided a method with which the other could be treated as an equal human being. For Van Lennep, the encounter had now become, following on from Buytendijk, the psychological method par excellence.[30]

For Langeveld, too, it was a principle that the methods of psychology should be assessed first on their anthropological and ethical merits. In his view, for example, statistical techniques were at most "nonessential aids." They deny the uniqueness of the person, which is objectionable on moral rather than epistemological grounds. According to Langeveld people can be influenced in such a way that statistics can be applied to them *epistemologically* speaking. However, that would have serious consequences for the essence of humanity: "then the entire human being rebels; individual and socio-pathological phenomena reveal themselves as ominous symptoms of a usurpation. Ignoring the singularity of the object, it is evoked in a terrible form, or . . . it is killed."[31]

That mainly moral standards were involved in the Utrecht protests against empirical-analytical thinking is also illustrated by Langeveld's strongest objections to De Groot's methodology. Prediction was frowned upon not so much because it was associated with generalization and therefore was not applicable for statements on individual people. As shown in Chapter 2, De Groot presented the prediction idea explicitly in combination with the recognition of individual uniqueness. However, Langeveld rejected this idea for the psychodiagnostic study of individuals in particular. To his mind, assuming the predictability of individual behavior denied people's freedom and responsibility. He believed that a psychologist had "to help see what is possible and useful" and that "is something essentially different than pronouncing hypothetical judgments."[32] The goal of psychodiagnostics should be to *educate* people rather than to ascertain facts about them. According to Langeveld it was "completely irresponsible and scientifically incorrect, to not orient a profession, in this case that of 'psychodiagnostician,' in terms of training and practice, to that which is the essential act of the profession," and this act is "*not* the so-called psychodiagnostic research . . . but the concrete influencing of our fellowman."[33]

As I already stated, although there were unmistakable differences in the thinking among the School's members, it can be said that with regard to the basic principles of psychology they nevertheless distinguished themselves as a group from the other "nonphenomenological" psychologists. The methodological unity of the Utrecht School existed in the agreement on the precedence of anthropology and morality above method. The goal of psychology was to lead people to full humanity. Full humanity was defined as the capacity to choose freely for the good, without the duress of guilt and punishment through internalization of morality. In order to promote this goal, a psychologist first had to develop a capacity for disinterested encounter with his fellowman. The latter could be imparted to students only through education to an inner refinement and not by learning rules and techniques for empirical research.

Phenomenology and personalism

The name "phenomenological psychology" could also lead one to overlook the Utrecht School's methodological principles. Anyone interpreting the concept of phenomenology to mean a method of understanding
phenomena in their uniqueness, inspired by the German philosopher
Edmund Husserl, expects rules in the work of the School for that purpose. In that case, too, one is looking for the sort of alternative that is
not really offered by the Utrecht School. Were the Utrecht psychologists
wrong to call their psychology phenomenological? The answer to that
question depends in the first place on whether or not one indeed wants to
reserve this concept for a Husserlian view.

In early twentieth-century philosophy, *personalism* appeared alongside
phenomenology. Personalist thinkers show similarities with philosophers
who are considered phenomenologists, but personalism could also be
regarded as an independent movement. The central positions of personalism offer more clarification of the internal coherence of the thinking and
goals of the Utrecht psychologists. In terms of present-day distinctions,
in personalism, too, anthropological and ethical postulates preceded epistemological ones.

As with phenomenology, many different philosophers and schools are
ascribed to personalism. However, twentieth-century thinkers who are
referred to as personalistic by themselves or others generally regard the
German philosopher Max Scheler as their predecessor. For the Utrecht
School, too, as will be seen, it was precisely Scheler's ideas that were of
major importance. Personalistic thought is therefore now explained on
the basis of a number of important characteristics and themes from
Scheler's philosophy,[34] which offers a further clarification of the methodology of the Utrecht School.

Scheler was primarily interested in practical questions of a psychological and sociological nature. He saw the values of Christianity being lost
in the society of his time. In bourgeois capitalism, he argued, quantity
prevails above quality. Scheler sought a basis for the validity of Christian
values, but he rejected the idealism of the neo-Kantians, whose doctrine
of objective values left, to his mind, too little room for personal freedom
and responsibility. Husserl's philosophy offered him the means. Scheler
developed an epistomology, which he named a "phenomenological epistemology," but he did this so much in terms of his own insights that
Husserl himself did not regard him as a fellow phenomenologist.[35] For
Scheler phenomenology was more a *stance* than a method. Falling back
on Augustine's Christian philosophy of participation, he espoused the
"knowledge of love," whereby "one opens oneself to the phenomena";

the phenomenological stance is characterized by the loving participation in being. Real objectivity, as Scheler put it, is possible only through "disinterested devotion." Along this emotional-phenomenological route, in the first place, objectively valid *values* could be recognized. Scheler developed this new foundation for the absolute hierarchy of values on the basis of the knowledge from love in his most important work, *Formalism in Ethics and Non-formal Ethics of Value (Der Formalismus in der Ethik und die Materiale Wertethik,* 1913),[36] with the subtitle *New Attempt to Found an Ethical Personalism (Neuer Versuch der Grundlegung eines Ethischen Personalismus).* A key idea in this is that people themselves sense the absolute hierarchy of values, as long as their pure feeling is not damaged by "resentment" or the psychic self-poisoning caused by capitalist society.

However, not just the system of values, but also a *human subject* can only be known by means of the loving devotion. The knowledge of positivism is not suitable for this. According to Scheler, this is based on an emotion other than that of loving devotion. The positivist is not interested in what the world really is; he wants to know how it is put together so that he can change it for his own purposes. He focuses on the verifiable and manipulable. This produces no knowledge from love but from the attempt to govern and achieve (*Herrschafts-oder Leistungswissen*). The general laws of the positivists, achieved through induction and deduction, express only the sum of separate entities. Positivist knowledge may be suitable for pragmatic purposes but not for the knowledge of essentials.

But knowledge of essentials is not just given to everybody. It makes *moral* demands on the knowing subject. There is an essential connection (*Wesenszusammenhang*) between moral requirements and the capacity to know: "a connection in which the moral and the theoretical world are attached to each other – as with clamps – forever."[37] In his essay, "On the Essence of Philosophy and the Moral Condition of Philosophical Knowledge" (*Vom Wesen der Philosophie und der Moralischen Bedingung des Philosophischen Erkennens*), Scheler discussed the relevant conditions.[38] A first requirement is the ability to love, the second requirement concerns the capacity for humility, and the third the capacity to repress bodily drives.[39]

Also necessary for achieving knowledge is *Bildung*. This concept, important for Scheler, cannot be simply translated as "education." It refers to an inner control of impulses and emotions and to the first two requirements for knowledge: the capacity for love and humility. Perhaps the best translation is "cultivation."

In Buytendijk's work from the Utrecht School period, similarities can be found with the work of many other philosophers. Buytendijk's spiritual forebears could be said to include numerous philosophers – Husserl, Heidegger, Binswanger, Guardini, Plessner, Merleau-Ponty, Straus, Von Weizsäcker, Plato, Aristotle, Augustine, Goethe. Buytendijk was extremely well read, maintained many contacts, and freely quoted anything from international philosophy that was useful to him in what he wanted to communicate.[40] Nevertheless, in his writings there is a striking affinity to Scheler's work. As an expert on Buytendijk writes, he "was profoundly impressed by the person and work of Scheler, large parts of which were literally adopted into his own work."[41]

The resistance to domination-seeking positivism, the relation between an objective moral, disinterested love and real knowledge, and the necessary education to love are all themes that were constantly in the forefront of Buytendijk's work. Scheler and Buytendijk were also good friends until Scheler's death in 1928. They had met during the period when Buytendijk was working at the Free University of Amsterdam.[42] At Buytendijk's invitation, Scheler gave guest lectures in Amsterdam in the 1920s, and Buytendijk taught with Scheler in the German city of Cologne. Scheler, who himself had twice joined the Catholic Church (and twice left it), guided Buytendijk not only to personalistic thought, but also toward Catholicism. In 1937 Buytendijk, who had been born into the Reformed Church and later became a Calvinist, was also received into the Catholic Church.

However, Buytendijk is not the only one who can be considered a personalistic thinker. In the period when he sought inspiration in Scheler's ideas, the much younger Langeveld was also receiving his education in the personalistic sense. Langeveld's spiritual father was the famous Dutch personalist Philip A. Kohnstamm. Under the name biblical personalism, Kohnstamm had also developed an axiology in which the active person is central, after he came to the conclusion that the axiology of the neo-Kantians left too little room for individual freedom and responsibility.[43] Although thermodynamics was originally Kohnstamm's major field, in 1917, after converting from Judaism to Protestantism, he chose pedagogics as his career. He became director of the Nutsseminarium for pedagogy in Amsterdam, which he helped to found in 1919. In 1932 he became professor of pedagogy at the University of Utrecht and in 1938 at the University of Amsterdam also.[44]

From his schooldays onward, Langeveld had been regarded as one of the family in Kohnstamm's home. In the 1920s when Langeveld announced that he wanted to concentrate on psychology, on Kohnstamm's advice he consciously did not go to Heymans in Groningen or Roels in Utrecht, but rather to Hamburg, where fellow personalist William Stern was a professor. He also studied with Theodor Litt in Leipzig and in this period he got to know Helmuth Plessner and Eduard Spranger. In the 1930s Langeveld worked with Kohnstamm in the Nutsseminarium for pedagogy. In 1939 he succeeded him as professor of education at the University of Utrecht, where his teaching commitment was expanded to include developmental psychology. In the war years, Langeveld replaced Kohnstamm, who was forced to go into hiding, as professor of education at the University of Amsterdam.[45] They were close friends until Kohnstamm's death in 1951.[46]

Van Lennep, finally, was known not only by Buytendijk and Langeveld through his Foundation for Psychotechnics, but he also shared an interest in personalism. He had studied theology with the "phenomenological theologist" Gerardus van der Leeuw in Groningen. He knew Kohnstamm, who appreciated his working methods in the Foundation. He was also a friend of Plessner, whose book, *Lachen und Weinen,* was exhaustively studied in the Foundation. During the war Van Lennep's contact with Plessner deepened when he got his hands on a copy of *L'Etre et le Néant* by the French existentialist Jean Paul Sartre; they even read the book together at Plessner's hiding place.[47] Immediately after the war, as the story goes, Van Lennep managed to track down Sartre "via a Parisian bookseller" in the café *Les Deux Magots* and in 1946 Sartre and his companion and kindred spirit Simone de Beauvouir came to the Netherlands at Van Lennep's invitation. This interest in Sartre and de Beauvoir requires further explanation. The *existentialism* to which they belonged – along with other French philosophers like Gabriel Marcel and Maurice Merleau-Ponty – also harked back in many respects to Scheler. But the thought of Sartre and de Beauvoir was only partially acceptable to the Utrecht School and the Dutch Christian public in general. They were mainly appreciated for the personalistic elements that were found in their work, but not for those points on which they deviated from it and that were essential for their philosophy. In particular the "absolute freedom" that was ascribed to man in Sartrean thought went much too far for the Dutch personalists. Sartre and de Beauvoir *disputed* the a priori character of values. In their existentialism man cannot depend on any "meaning" of existence, on any fixed morality. On the contrary, he constantly has to make up his mind.

It was exactly on this point that existentialism was not accepted by the members of the Utrecht School. For example, Buytendijk wrote a book entitled *The Woman (De Vrouw)* as a correction[48] to de Beauvoir's book, *Le Deuxième Sexe.*[49] Although he subtitled his book *An Existential-*

Figure 8. J. P. Sartre, S. de Beauvoir, and D. J. van Lennep in 1946 (photo courtesy of W. Wijga).

Psychological study, he made it very clear that he had no Sartrean existentialism in mind. In de Beauvoir's book he found a lack of the human soul:

> Reflection upon the inner is according to her meaningless, because man only is what he does, because nothing can be therefore met in the so-called inner which has not been thought, done etcetera by himself. Man therefore finds nothing in himself which does not belong to his being in the world. There is no encounter possible with his essence which defines and demarcates his existence, which precedes existence.[50]

On the contrary, according to Buytendijk, it must be "recognized" that

> each human existential project is naturally designed in an inevitable turning towards an order of *objective* absolute prescriptive values. The fundamental mistake of Sartrean existentialism is in the denial of this, in so far as it defines human freedom not only as a being-free from every subjective essential predetermination, but also as a command to maintain freedom in the face of every objective absolute prescriptive value.[51]

Although some members of the Utrecht School based themselves on existentialist philosophers – their attention to "the situation," "the landscape," "ambiguity," and the "body-subject" stemmed from Merleau-Ponty, for example[52] – this existentialism did not bring about any profound change in the original thought of the core group.[53]

The answer to the question of whether or not the Utrecht psychologists were incorrect in calling themselves phenomenological psychologists depends, as stated earlier, on whether one wants to make a distinction between personalistic thought and phenomenological thought. There are similarities between personalistic psychologists that are not shared with – other – phenomenological psychologists. The psychology of the Utrecht School also shares these traits. Unmistakably personalistic is the internal intertwining of an objective theory of values with a specific epistemology; as Scheler expressed it, morality and methodology are considered to be linked together "as with clamps." Research styles are interpreted as positions that are good or bad in the moral sense. The capacity to really know is also dependent on moral requirements from the knowing subject. This capacity is transmitted by education or, to put it better, *Bildung,* or cultivation.

The designation "personalistic psychology" would therefore seem to be at least as accurate as "phenomenological psychology." In a second interview in 1987, I asked Langeveld, who in an earlier interview had used only the term phenomenology, whether the term "personalistic psychology" would not have been appropriate with regard to the Utrecht approach. Langeveld answered that that would have indeed been so "if a certain recognition of that term was realized." However, according to Langeveld, in the Netherlands of that time personalism was "a concept you had to be careful with."[54] Langeveld's remark offers points of departure for the further study of how it came about that the phenomenological psychology of the Utrecht School could thrive in the postwar years, but then quickly lose its following.

The leading elite of personalistic socialism

From the very beginning, personalism was not just a philosophical school, but also a *political and sociocultural movement*. In the Netherlands, it flourished during the Second World War and found a strong following in some circles after the war, but at the same time it soon became controversial. This political variety of personalism, *personalistic socialism,* is the broader context within which Utrecht phenomenological psychology was established.

A brief description of its ideals will further clarify Langeveld's policy in the postwar reconstruction of Utrecht psychology. Personalistic socialism was opposed not only to the capitalist exploitation of the proletariat, but also to Marxism. Whereas capitalism was accused of regarding man as a separate individual, Marxism was accused of seeing him as nothing more than a part of a whole. According to personalistic socialism, man is not merely the object of history and a product of economic circumstances, but must in the first place be regarded as a being that realizes values and ascribes meaning. The basic principle must be the personal responsibility of each individual for the good and evil of the entire community. Community, solidarity, and responsibility were the key concepts that personalistic socialism applied in its battle to elevate and emancipate the working population. Objectively valid values, not historical determinism, were the indisputable basis of this socialism.

The task of a social vanguard, then, is not to work for the realization of an objective development of history, as in Marxism. Rather, that task is to clear the way for everyone to realize transcendental values through popular education. Personalistic socialism must be brought about by a *change in mentality* instead of by revolution. This is what Scheler actively sought and what Kohnstamm tried to achieve through his program of "Popular Education." A group of French-Catholic philosophers around the journal *Esprit* also presented a Christian-socialist creed in the 1930s. These philosophers, Maritain, Mounier, and De Rougemont, became popular in Christian personalistic circles because of their message that the point of application for changes must be found not in socioeconomic factors, but in personal-psychological ones. The name personalistic socialism was borrowed from them.

The leading figure of Dutch personalistic socialism was Willem Banning. In 1931 this reformed preacher and social democrat founded the so-called *Arbeidersgemeenschap* (Workers' Community), with the aim, as he put it, of

> planting or helping to grow the seeds of religious life in the working masses, and on the other hand to preach socialism as an acceptable doctrine in the Christian section of the population. The movement therefore opposes both an egotistical or socioconservative religion and a materialistic-atheistic philosophy. All this means: no political program, no new party, church or sect. It means a principle of living, a source to draw on, a power to reinforce one's own power.[55]

With financial help from the socialist novelist Henriette Roland Holst, the Workers' Community in Bentveld, near Amsterdam, acquired a conference center that Banning became director of. Lectures were given, courses organized, and plays by Roland Holst performed. From the beginning, the young Langeveld was a regular participant in the activities in Bentveld. He took part in the discussions and cultural activities, he knew Banning well, and became a friend of Roland Holst.[56]

The Second World War brought the activities of the Workers' Community to an end. The buildings were confiscated by the Germans and shortly afterward Banning and a number of his kindred spirits ended up in a camp for hostages, which the Germans set up in the South Dutch village of Sint Michielsgestel. There, leading Dutchmen – politicians, professors, writers – had to guarantee with their lives, as the Nazis put it, "the maintenance of peace and order in Dutch society."[57]

At the camp there was the constant threat of execution and some hostages were murdered. Nevertheless, conditions in this camp for the rest were less trying than the conditions in the other concentration camps. The Dutch business sector ensured that there was good food, and there was space for cultural events, training programs, and even political deliberation. Thus, in the Sint Michielsgestel camp Dutch personalistic social thought could be further developed. Through some members of the Bentveld Workers' Community the ideas of the aforementioned French *Esprit* group were introduced. On the basis of these ideas, the participants argued that, among other things, after the war the borders between the various religious and nonreligious political parties must be broken down. There should be no place in the postwar Netherlands for the "pillarization" or segmentation of society described in Chapter 2. People must unite for moral and economic reconstruction. They hoped for an "encounter of groups," which could now be seen in the camp, even though it was enforced.

Not only social segmentation but also *parliamentary democracy* was raised for discussion by this elite group of Dutch society. The idea was that democracy must be curtailed after the war in favor of a government of intellectuals that would provide leadership in changing popular mentality. Parliamentary democracy could be reintroduced after the Dutch people had been so educated that they could make choices democratically. The elite therefore should be concerned not only with administrative affairs, but also be interested in questions about human personality. Those who rule people, the idea was, must have *psychological* insight. They must know the conditions of humanity and know which rules must be followed to guarantee the meaning of human life. A group formed around Banning and the politician Willem Schermerhorn who took upon themselves the task of becoming the leading elite after the war.

Three weeks after liberation the Michielsgestel core group founded the Dutch People's Movement (*Nederlandse Volksbeweging*, hereafter NVB) with Schermerhorn and Banning as chairmen.[58] An appeal was made to the Dutch people to realize the renewed socialism. The text of this appeal had previously circulated in a limited group and was signed by 450 sympathizers. The most important program points included: "development of the human personality," "renewal and safeguarding of family life," "elevation of work in an organized socio-economic system," and "reinforcement of the national community." In a political section a "strong and decisive government authority" was advocated, along with "more autonomous powers for the government," and "strengthening of the position of the prime minister."

The archives of the NVB, which have been preserved in the International Institute for Social History in Amsterdam, contain this appeal and the list of the first 450 signatories. A number of the people already mentioned in this chapter in relation to the Utrecht School appear to have belonged to this very first small group of NVB supporters. The list includes, for example, Langeveld's spiritual father Kohnstamm, as well as Professor Pompe, who belonged to the Utrecht School in criminology and was rector of Utrecht University in 1946–7. In that capacity Pompe was helpful to Langeveld in realizing his plans. Rutten, the leading Nijmegen psychologist who helped Langeveld in the Buytendijk and Van Lennep appointments, signed the appeal, as did Kruijt, the sociologist who gave a subsidiary course in sociology to the psychology students of the Utrecht School.[59] Kohnstamm and Pompe in particular were strongly drawn to the movement. At a meeting in 1945 announced and intended for "a limited circle of people providing spiritual leadership," Kohnstamm discussed personalistic socialism from a Calvinist-Christian perspective, and Pompe did the same from the Catholic viewpoint.[60]

Langeveld's name does not appear on the list. The *possibility of membership* was not open to him at the time the list was being sent around; people who were suspected of collaborating with the occupying forces were not allowed to sign. Soon after liberation, an investigation was begun into Langeveld's position during the war years and in September 1945 he was suspended by the "Purging Committee."[61] However, this was quickly followed by many statements in his defense. Whereas some students had earlier declared that at the beginning of the war Langeveld had called noncooperation "treason," other students now gathered signatures on his behalf. Although some colleagues accused Langeveld of deliberately undermining anti-German actions, Kohnstamm testified that Langeveld had helped him to get identity papers that made it possible for him to go underground. In November the purging measure was lifted.[62]

Nevertheless, Langeveld certainly had clear sympathies with personalistic socialism. His visits to Banning's Workers' Community showed that already. Ironically enough the purging measure, which made it impossible for him to sign the NVB manifesto if he had so wished, was also the consequence of a typical personalistic-socialist activity. In the 1930s Langeveld had already begun to work on behalf of the plan for a compulsory Work Service, which originated in Germany, and was also put forth in the Netherlands by the socially and politically active psychologist J. E. de Quay. The idea of this was that "young men of a particular age, irrespective of rank or class" should all be compulsorily included in a workers' community for six months, where they would be educated according to the principle "one for all, all for one." Here, community feeling and discipline could be instilled in the young men. However, this idea was also popular in Nazi Germany, and after the occupying forces had taken over the organization, Langeveld did not withdraw immediately.[63]

After the war, Langeveld also called himself a personalistic socialist[64] and in 1945 he offered the NVB a text for publication with the title "Conscience Co-Governs" (*Het Geweten Regeert Mee*).[65] When a new party, the Labor Party (*Partij van de Arbeid*, hereafter PvdA) was founded in 1946 under Banning's leadership, this party was joined by NVB members who had political aspirations; Langeveld became a member, too. As a personalistic socialist he gave lectures throughout the country for the PvdA, he led training meetings, and he wrote on educational policy for the party journal, *Socialism and Democracy (Socialisme en Democratie)*, edited by Banning.

Although it is true that the first postwar government (the so-called cabinet of reconstruction and renewal, which convened from June 1945 to July 1946) included a remarkable number of NVB members, the ideas of the NVB failed to catch on in broader circles. The NVB remained the interest of a small elite.[66] Nor did the goal of breaking up social segmentation amount to much. It was not possible to create a unified movement from the various Dutch parties. In 1944 the bishops had made it clear to the Catholic population that the organization of the Catholic groupings should be restored "now, without delay" and that people should not associate with non-Catholics. This led to the rise, on the Protestant side, even in circles favoring the breakdown of social borders, of a strong anti-Catholicism.[67]

After the foundation of the PvdA, the NVB lost a large number of its members. The political part of the personalistic quest was now left to Banning and the PvdA. Under Kohnstamm's guidance, the NVB remained active for several years as "a movement for spiritual encounter," but in 1951 it was also abolished as such. However, in 1949 Kohnstamm founded the "Discussion Center" at Driebergen as a "meeting place for all denominations." In this center, which still exists, the personalistic mentality was further encouraged, along the same lines as the still existing Workers' Community.

The Discussion Center also attempted to institutionalize the dissemination of the personalistic mentality in less noticeable places. For example, Kohnstamm had argued for the addition of a department of social psychology and social pedagogy at the University of Amsterdam "for the benefit of the cultivation of leaders." Here, against the wishes of Kohnstamm and his allies, the university appointed Tonko T. ten Have from Groningen as professor, who as a student of the empirical-analytical-oriented Groningen psychologists Heymans and Brugmans, did not make the new field the social discipline that the initiators had had in mind.[68] The postwar reestablishment of the psychology department at the University of Utrecht is an example of a successful institutionalization of the personalistic quest, though. In the next section I will show that in reestablishing the department Langeveld was guided by the personalistic-socialist ideals on the role of the university and the academic in society.

The responsible university

In the hostage camp Banning had put his ideals into writing, which after the war appeared under the title, *The Dawn of Tomorrow: Outline of a Personalistic Socialism: Guideline for Renewal of our National Life*.[69] In his book, Banning discussed the idea that universities must become aware of their role of producing the leading elite of society. Therefore, he contended, it should no longer be possible for the universities to restrict themselves to the practice of pure science:

> There is perhaps no institution where the need for a radical switch from neutrality to responsibility is illustrated more clearly than in our Higher Education. . . . At this moment in history we cannot be content with clever heads alone, it is above all a question of conscious spirits and devout hearts. In other words, in this new age with new requirements, the University, which up to the present has found its purpose in the education of its students for the independent, self-reliant practice of science, and training for particular functions and posts, can no longer be content with this. It is above all the University, the source of free intellectual life, nourisher of all those who pass on the intellectual heritage to the people in the broader layers of society, which cannot lag behind, but must consciously and proudly forge ahead where clear choice is demanded. . . . The education of the layer of intellectual leadership of our people must take place on the basis of those values and norms contained in Christendom and Humanism.[70]

These personalistic ideas on the role of the university graduate in society can almost literally be found in Langeveld's postwar arguments. He, too, expounded the idea that the university graduate must be a leading citizen and the task of the university is to educate him to meet this goal. He spoke of "the educational task which we have inescapably acquired" and "which we can no longer avoid with the claim that the university is only concerned with the practice of pure science."[71] "As representatives of science and of society," he believed,

> the teachers must help their students to found an inner sanctum in themselves in which the value of disinterested truthfulness is a living presence. It is in the confrontation with "L'angoisse et peut-être l'agonie de la personne humaine" that we want to watch over the education of the student and it is in the light of that task that we dare to demand that the Government take charge of the abolition of all the impermissible situations in our current university problems.[72]

At a 1948 conference of the International Society of University Professors and Lecturers, Langeveld gave a paper under the title, *The University's Part in the Education of the Citizen*. Here, too, he stressed the task of the university in making a contribution through the education of students to "the integrating values of human life." Above all, the university should form the spirits of the students in such a way that they do not merely become people who know a lot and can do a lot, but people with "a deep and sincere respect for human dignity": "In that case they will be *leading* citizens, leading not only by their knowledge, but more than that by their insight in the things of life."[73]

At first the appointments of both Buytendijk and Van Lennep as professors in psychology did not seem very obvious, but when they are seen in terms of Langeveld's social ideals these choices are quite understandable. It was not so much the intention in Utrecht to realize a specialized psychology training, but rather to found a university training institute where the cultivation ideal could be given shape and where norm and empiricism were not unlinked in the practice of science. This goal required a scientific staff that was characterized primarily by "general wisdom and inner refinement."

It could also be said that the ideal of breaking down denominational borders was actually realized in Utrecht psychology. Not only were people from very different denominations appointed to the same group (which was not at all common in the Netherlands of the 1950s),[74] but the barriers between the scientific disciplines were also broken down. That Buytendijk was not a psychologist by training and that Van Lennep was hardly recognized as a colleague by many other psychologists was absolutely unimportant. What was important was that they were known as people who could transmit the values of personalistic thought. The more peripheral members of the group, such as Rümke, Van den Berg, Pompe, and Kruijt, were not professional psychologists either, but they did think in terms of a similar image of man and society.

From this perspective, Buytendijk was not an odd man out in psychology but rather the person par excellence to fill the Roels vacancy. In contrast to some of his good friends and acquaintances he was not a personalistic socialist in the political sense,[75] but more important was that for many years he had been promoting the cultivation idea with regard to the academic world. Already in 1922, the first article by Buytendijk (inspired by Scheler) appeared in the pedagogic field, "A Consideration of Some Contemporary Pedagogical Problems" *(Beschouwingen over enkele Moderne Opvoedkundige Problemen)*, a version of which later appeared as *Erziehung zur Demut (Education to Humility)*.[76] That piece was written in the period when Buytendijk was defending in Amsterdam Calvinist circles an "anthropological medicine," in which the patient's individual responsibility for his bodily condition was a central tenet (see Chapter 4). Buytendijk's lessons on this at the beginning of the 1920s were taken by Langeveld as a student.[77] In newspaper articles before the war Buytendijk had already regularly expressed his concern at the loss of the cultivation ideal in the universities, which he blamed on a positivist tendency. In those newspapers he wrote not only pieces with titles like *Why Do We Play Yoyo?* and *Do Animals Feel Homesickness?* but also, for example, on *The Task of the University: The Need to Link Up with Life*. Moreover, in 1939 Buytendijk had already showed his complete agreement with the ideas of the French-Catholic personalistic socialist Maritain.[78] He did this on the occasion of the third anniversary of the Nijmegen students' union, for which Buytendijk was invited as a speaker, along with Gabriel Marcel and Henry Daniel-Rops (both also French-Catholic personalists). In such lectures Scheler's idea of the *Bildungswissen* was an important principle.

Finally, in "The Ideals of the Ideal Student" *(De Idealen van den Idealen Student)*, delivered in 1945 to the Utrecht and Groningen Catholic students' unions, Buytendijk impressed on the students the need to strive for personal cultivation:

If you make this striving the motor of your university studies then you will come to belong to the intellectual aristocracy which in sincere humility bows before Him who is the Lord of all nature and history. . . . Then you will one day be able to fulfill with honor the task which by virtue of your cultivation devolves upon you and which lays upon you the high constraint to be not the slaves but the leaders of the Zeitgeist.[79]

Unlike Buytendijk and Langeveld, Van Lennep had not moved much in university circles before his appointment in 1949. The education of students was not one of his concerns, let alone the notion of creating a leading elite out of them. However, in his Foundation for Psychotechnics he applied himself to the staff selection for higher functions. Furthermore, as described in Chapter 2, he did not see psychotechnical research merely as a means of selecting staff, but above all as a means to help people give shape to their existence. His work was not explicitly based on a strict division between norm and fact either. In this respect his methods and ideas on psychotechnics deviated from those of psychotechnicians like Prak and his colleague De Groot, whose specialty was (merely) in supplying information on the job performance to be expected.

Van Lennep therefore was something of a popular educator as well. When the professorship in applied psychology had to be filled, he fitted in better than other practicing psychologists did with the plan to have the new psychology department play a role as the educational institute for a leading elite. Langeveld and Buytendijk surely had reasons to maintain, against the express wish of the faculty, their choice of Van Lennep over De Groot (or any of the other seven candidates who had been recommended by the psychology professors of the other universities).[80]

In the Netherlands of that time the concept of personalism was too political and controversial to make it eligible for the name of the new psychology. But "phenomenological psychology" was acceptable. And so in the 1950s the Utrecht psychologists, and the circle of like-minded psychiatrists, medical doctors, educationalists, and philosophers around them, worked under this name to establish a psychology in which the desired "switch from neutrality to responsibility" would be given shape. As has already been described: The students were actively made aware of the fact that they had been called to a special responsibility; with the epistemology of the "encounter" an alternative was developed to the neutral and control-oriented gathering of facts that was characteristic of "the natural sciences-based psychology."

The phenomenological psychology of the Utrecht School was not a detour from a straight road toward quantitative and predictive contemporary psychology. It was, on the contrary, a way that was supposed to lead to ideals of a completely different composition. Moreover, these ideals were not autonomous, but were part of a conglomerate of ideas on correct societal relations and the correct division of political power. In the phenomenological psychology of the Utrecht School, both a scientific and social identity of psychology was expressed.

Impending changes

For a short while it appeared that psychology at *all* Dutch universities would undergo the "switch from neutrality to responsibility." The theologian Gerard van der Leeuw was appointed Minister of Education, Art, and Sciences in the first postwar Cabinet of Renewal and Reconstruction. Van der Leeuw was an enthusiastic supporter of the NVB, and in his book on the Netherlands of the future he also emphasized the educational task of the university.[81] In 1946 this minister set up a committee to prepare a piece of legislation on training in psychology, which was to replace the 1941 regulation introduced by the German occupying forces.[82] This "Committee on the Training and Title of Psychologists" was supposed to submit a proposal on protecting the title of psychologist. An earlier committee, which had been delegated by the Dutch Association of Practicing Psychologists (the NIPP) to the Ministry of Social Affairs, was then dissolved and partly absorbed into the Committee on Training and Title.

Problems soon arose. The psychologists from the University of Amsterdam noticed that they were not represented on the committee. The Amsterdam professor Géza Révész made various attempts with both the minister and his successor (Jos J. Gielen, also an NVB supporter) and the committee itself to become a member, but he was excluded.[83] It was argued that (the educationalist) Kohnstamm represented the Amsterdam university and that the group was already big enough.

Further examination of the composition of the committee however, shows that it was dominated by personalistic thinkers. The chairman was F. Sassen, secretary general of the Ministry, a Catholic philosopher, and member of the NVB. Apart from Kohnstamm the other members were Langeveld, Buytendijk, Van Lennep, Rümke, the Catholic philosopher and Langeveld friend Father C. Van den Berg, and, until he became Minister, also Rutten. Nor could much be expected in terms of "Amsterdam standpoints" from the other members: Apart from Rümke, Brugmans from Groningen and the Protestant psychologist and preacher Waterink were taken on from the NIPP committee. An added member was W. J. P. Willems, the Catholic psychologist who worked for the State Mines.

When the committee published its conclusions in 1949, the report of course met with some protest from the Amsterdam psychologists. Professor Duijker wrote about it in the *Dutch Journal of Psychology (NTP),*[84] and De Groot devoted a couple in the 1949 inaugural lecture marking the commencement of his professorate in applied psychology. In De Groot's opinion, the "empirical-scientific viewpoint comes off badly among the pedagogy, philosophy, and ethics."[85] He argued for statistics as a compulsory subject.[86] He also protested against the committee's intention to give educationalists extensive powers in psychological practice:

> Because in spite of all similarity, there is a difference in principle, comprehensive and difficult to bridge, between psychologist and educationalist. As a matter of principle and in spite of all practice, the first one ultimately does not want to know what people are like, for after ,all, that is the object of his research. The other, on the other hand, even knows what a human being *should* be like.[87]

It was not only the Amsterdam psychologists who reacted. The younger generation of the NIPP informed Rutten by telegram that they intended to submit a counterproposal, which they did in 1950. On 28 May 1952 the legal regulation with regard to psychology was changed by Royal Decree. This deviated so much from the recommendations of the Sassen Committee – statistics became a compulsory subject, for example, but not pedagogy – that in a letter to Rutten, Langeveld strongly protested against the "weakening of the education of students."[88]

The personalistic psychologists had lost this battle. The design of psychology training in accordance with their ideals was limited to their own Utrecht training (and also had to stay within the limits of the Decree, which were actually still broad enough). As already described, that was quite successful: In the 1950s Utrecht phenomenological psychology was in its heyday. In 1953 the subversive student magazine *Parasol* was certainly too early in publishing the following poem on Buytendijk, whose name in Dutch literally means "Outerdike":

> Help O Lord, now the Outerdikes break down
> all around the ancient Utrecht town
> the doctrine of the Fathers is rejected
> and reverence in the world neglected.[89]

Nevertheless, during this productive period, changes were taking place within the group that would later continue in the Utrecht psychology of the 1960s. In the course of the 1950s, Van Lennep, who remained closely associated with the practice of staff selection – where predictive thinking had most of its followers – and who from the very beginning was less inclined than Buytendijk and Langeveld to stick to principles, lost a number of his former methodological convictions. On occasion he even openly showed his approval of the growing influence of empirical-analytical thought in Dutch test psychology. For example, in a 1957 lecture he distanced himself from what he had vigorously defended in his 1949 book on psychotechnics.[90] He said that he now ascribed more value to the validity of tests than to the "inner evidence of intuition" and he called the idea that every psychologist is his own best instrument "misleading." Of projection tests, he now took the "Amsterdam" view: They have a function in the formation of hypotheses.[91]

It was, however, the younger generation who most noticeably went beyond the borders of the phenomenological program. Kouwer, who had also studied elsewhere and had been appointed to the Utrecht psychology department in 1946 before Buytendijk and Van Lennep arrived, soon turned out to have unmistakably deviant viewpoints on phenomenology. In Kouwer's view, science and thus psychology could not be phenomenological. Insofar as someone practices science "he explicitly posits the necessity of seeing nature, the world, as that which can be studied and analyzed. Nature is subjected, it is placed in the bonds of laws and rules, structured into facts and objects."[92] Phenomenology therefore had a role *beside* science: As "modern magic" it offered a ritual way to experience the essence of phenomena. The experience of essences, from Kouwer's point of view, was of a nonrational nature and therefore phenomenology could not be called science.

In terms of philosophical anthropology Kouwer was more a follower of Sartre than of Scheler. For Kouwer, too, man has no "meaning" or "soul," but "is" only that which he chooses to do. When he became professor of psychology at Groningen in 1955, he gave an inaugural lecture on the social role of applied psychology. Without saying explicitly that he had the Utrecht way of thinking in mind, he attacked the idea of the psychologist's task as "modern priest" or "pseudo-doctor":

> What norm should the psychologist aim at in his practice? Not in any case, at "health." It is true that it has become the fashion to speak of "mental health" but that concept has hardly anything to do with what a doctor understands by health. The mental, the psychic, is not an extra organ of our body, there are no "healthy" or "normal" workings of the mind imaginable, and it is impossible to determine once and for all, for everyone, how a healthy mental life should proceed.[93]

Some say that after this relations between Kouwer and most of his Utrecht ex-colleagues grew cool, and that is not unlikely, especially when seen from the Utrecht viewpoint.[94] Although Kouwer did not say it in so many words, the Utrecht thinking was attacked at its core. The

statements on mental health care quoted in the previous paragraph literally contest the legitimacy of Buytendijk's very forceful attempts to set
up – in the Catholic Netherlands – a guiding mental health care. Buytendijk promoted the idea that mental health was a precondition for the
ability to keep to the rules of religion and morality. In the case of
transgression, punishment by the church would have to make way for
treatment by psychologists and social workers. The goal of such treatment must be to enable individuals concerned to rediscover mental freedom so that they could turn on their own initiative toward "the true and
the good."[95]

The idea of the psychologist's task, which Kouwer attacked in his
lecture, was also clearly the view Langeveld advocated. Although
Kouwer discussed and agreed with Langeveld's opinion that a question
about suitability for a particular career is often a veiled request for
guidance in life, in contrast to Langeveld, Kouwer did not then proceed
to say that the psychologists should actually *give* this help. True to his
existentialist orientation, he warned his colleagues that on the contrary
they had nothing to offer in that area. A psychologist, as Kouwer put it,
could not give an individual being studied a reason for his existence, but
only the result of the tests. The task of the psychologist would be to give
the individual the existentialistic insight that he or she was not defined in
any way by these results. Responsibility for the future was completely in
the subject's own hands: "Man has not been left at the mercy of his
character, as if it were a destiny: He *possesses* a character, that is, he
has a character at his disposal and what this character must become is in
the future, not in the tests."[96]

Linschoten, who succeeded Buytendijk after his retirement in 1957,
also distanced himself from the Utrecht ideas. To begin with, it is true,
he had written a series of articles that fit well in the context of the
Utrecht School, and he had also dedicated his book on William James
"to my teacher Prof. Dr. F. J. J. Buytendijk,"[97] but he gradually developed the view that phenomenology and empirical psychology are two
activities that must be distinguished from each other. Linschoten now
allocated psychology the task of eliminating preconceptions by means of
standardization, quantification, and testing hypotheses. In his book, *Idols
of the Psychologist* (*Idolen van de Psycholoog,* 1964), which appeared
after his death and became famous in the Netherlands, he emphatically
advocated a value-free, empirical-analytical psychology. For phenomenology in the Husserlian sense, Linschoten also reserved a task outside
scientific psychology. Phenomenology was supposed to provide the critical reflection on the world as experienced that produces the scientific
questions and determines how they are asked.[98]

Buytendijk and Langeveld were profoundly disappointed and distressed that their most important student had strayed.[99] The fact that Linschoten, like Kouwer, continued to see a role for phenomenology alongside scientific psychology has led some to believe that this disappointment was actually without foundation, and that Linschoten, in fact, was not an apostate.[100] However, both Kouwer's and Linschoten's ideas meant a real break with the central tenets of Utrecht thinking. Their phenomenology was not phenomenology as Kouwer and Linschoten would now see it. The School's *program* was that scientific psychology and phenomenology were in fact identical. Furthermore the School promoted phenomenology in the personalistic sense, which implies a phenomenological psychology that does not shy from making clear moral statements on good and evil: "There is conceivably a difference between a factual and a normative judgment," Buytendijk once wrote, "but in anthropology (and therefore in psychology) these judgments cannot be separated."[101] That was precisely what Kouwer and Linschoten turned against. For them, science had to produce facts and not norms. On the crucial point of the choice in favor of impartiality in science, seen from the viewpoint of Utrecht phenomenology, it actually boiled down to the fact that Kouwer and Linschoten had gone over to the side of "natural scientific" psychology.

A new morality and new methods

In the course of the 1960s it happened: The outerdikes broke down all around the ancient Utrecht town as well as in the broader surroundings. Although in the 1940s and 50s little came of the personalistic-socialistic ideal of breaking through the "pillarization" of society, it began to get under way in the 1960s, without a broad "popular movement" aiming resolutely at this goal. The process has already been described at the end of Chapter 2: The economy took off, religious bonds became looser, and the population was better educated and more emancipated. At the same time, piety and the power of religious prohibitions were rapidly declining and interaction between people of different faiths was growing.

At first glance, the personalistic-socialistic ideals would seem to have been realized "by themselves." However, the broadening of people's mental horizons brought with it exactly what the churches had predicted and feared: The traditional fixed beliefs in the field of marriage, family life, sexuality, and religion were also thrown open to question. Thus, as the "pillars" began to falter, the moral truths defended by personalism began to lose their objective validity as well. The moral convictions that education and cure must lead to lost their cogency. In the 1950s, trading in guilt and atonement by the church for illness and cure by therapists had meant a liberation, but by the second half of the 1960s the therapeutic model had increasingly come to be perceived as being equally oppressive. The message that obedience to moral laws should take place out of free will was now exchanged for the lesson that moral rules *themselves* had to be left to the individual conscience.[102] The new morality became that every individual had to do – in consultation with others – what he or she thought was good. In this way people learned how to fit better into an increasingly individualized world where holding on to universally valid moral principles would obstruct the indispensable capacity to function flexibly in variable social contexts.[103]

Again, psychology was part of society. While a group within the Labor Party under the name "New Left" advocated a break with Banning's generation of personalistic socialists,[104] psychology students at the University of Utrecht began to call for different training. The catchphrase *Buytendijk buiten de Dijk!* ("Buytendijk outside the Dike") found its counterpart in *Ba Banning!* ("Poo Banning") in an Amsterdam student debating club.[105] The students wanted modern – and that then meant positivist – research and training. The idea of an academically trained leading elite no longer had much support. "Cultivation" and "humility" did not sound lofty and respectful anymore; they sounded old-fashioned and insincere.

As far as psychology was concerned, ideas like Kouwer's turned out to have already ushered in a new era. According to the ideal image of the profession that was now growing in the universities, the researcher and the test psychologist could no longer offer guidance and support, but must provide only "factual information" with which the subjects could or in fact had to make up their own minds. From now on the treating clinical psychologist could only "guide" the subject in achieving insight into the absolute freedom and responsibility of the individual. In this manner, "the lonely person, the person gone astray, the inhabitant of the wilderness of society from the bushman type to the gangster, . . ." whom Langeveld had wanted to put on the right track,[106] in this new era was led to accept, with the powerful urging of a changing but growing interest in mental health care, that "the" right track does not exist and that people had to go their own way.

The Utrecht subdepartment of psychology grew and split into an increasing number of specializations. New professors no longer presented themselves as phenomenologists, but in diverging ways aimed at "value-free knowledge" by empirical-analytical standards. For research, empirical-analytical methodology was further developed with the help of American methods and techniques. For therapy, the communication of the existential morality of absolute freedom was further developed with Rogers's counseling method and other "nondirective" techniques. While typical phenomenological issues were dropped, new questions on human cognition and capacity, which had not been on the agenda under the phenomenological regime, were now raised and studied.

The phenomenology of the Utrecht School appears to have been linked so much to a culture in which personalistic standards had a broad enough basis, that when that culture disappeared, it could not survive. To put it another way, the social identity that Utrecht postwar psychology expressed in its methodology was no longer adequate at the end of the 1960s. Psychology in Utrecht assumed a new identity, expressed in a new methodology.

At the end of the 1960s when a small Marxist-inspired democratization movement already began to proclaim the end of the positivist era, understandably that was hardly a comfort to the old guard of phenomenology, who had seen their ideals getting lost. Although it is true that this newest generation of Marxist "critical psychologists" made a fierce attack on positivist psychology that it accused of "concealment," "veiled essentialism," and "isolated rationality,"[107] and although it is true that this generation argued again in favor of a psychology based on explicit values, these values were not those of the former professors. Buytendijk, who turned eighty in 1967, was disappointed with the new methods and had withdrawn further and further from academic life. Langeveld, who retired in 1971, saw himself as a voice crying in the wilderness. In letters to Buytendijk he described the "soulless unculture" into which society had degenerated, and he pleaded that the struggle against "emptiness" and "mechanistic psychology" not be abandoned.[108]

4

The neurotic paradox of clinical psychology

If the Utrecht psychologists of the 1940s and 50s were not much taken with De Groot's methodological ideas, the same is not true of all Dutch academics. When De Groot gave his inaugural lecture, "The Object of Psychodiagnostics" *(Het Object van de Psychodiagnostiek),* at the University of Amsterdam in October 1950, Juda J. Groen, an internist at the Amsterdam Wilhelmina Hospital, was sitting in the hall. Groen never forgot how enthusiastically he had applauded De Groot's speech.[1] The idea that only observable behavior can be the object of scientific psychological research was one close to Groen's heart, as was the notion that the "empirical cycle" runs from observed and recorded behavior to predicted behavior. Back in the Wilhelmina Hospital he wrote a note of congratulations to De Groot and asked him whether he could recommend a psychologist who could apply the working method described. As a result, Johan T. Barendregt, a recently graduated psychologist, joined Groen's "Psychosomatic Working Group" on 1 January 1952.

Barendregt was appointed as clinical psychologist with his most important task being to develop methodologically sound psychodiagnostic means. He carried out this assignment for a number of years, and received his doctorate in 1954 (with De Groot as supervisor) on a study of the testing of hypotheses from psychosomatics. In 1961 he published in English a collection of experimental research reports on the validity of clinical assessment, and in 1962 he became a full professor at the University of Amsterdam.

Throughout his twenty years as a professor – Barendregt died in 1982 – his interest continued to be a combination of clinical psychology and methodology. Over time, however, his viewpoints evolved: In the 1950s he could still make self-assured statements on the victory of "scientific psychodiagnostics" over "prescientific thought," but in later years he gradually came to be less sure. His methodological standards always remained within the framework of the predictive empirical-analytical methodology. However, if Barendregt regarded a research project as methodologically correct, he believed it was not relevant for life outside of science and hence for clinical practice, but if he considered a project to be relevant, he criticized it for not being methodologically correct. With his characteristic irony he wondered at the end of the 1970s whether psychologists must really "ascribe so much reality to the idea of science, that you would rather run the risk of not getting round to the soul than not to all the phases of the empirical cycle." Perhaps methodological rules should just be regarded as abstractions of what really goes on in doing research. They are "fruitful for writing the cultural history of science," but not as commandments telling scientists what to do and what not to do. After all,

> if you have lost all hope because your predictions again have not come true in your last study . . . , then you can still come to the conclusion that methodology is more scientific than psychology. And look for all your footing in even more methodology and even less psychology. Then you have ended up in another cycle, that of the *neurotic paradox*. Because for some psychologists at least the rule applies that man is inclined to attach himself to a romantic image: which ruins his life (italics added).[2]

This chapter looks at methodology in Dutch clinical psychology, with Barendregt as the main character. Barendregt's individual history is not representative in every respect of the history of clinical psychology in the Netherlands. In the description of Barendregt's development there are naturally a number of purely individual facts and events. Sometimes Barendregt even is illustrative more through contrast than through identity. Besides, no Dutch psychologist has been as visibly thwarted by the "neurotic paradox" as Barendregt has. More than anyone else, according to those who have known him, he tended to take the methodological rules seriously and to their most extreme consequences, only then to become angry and full of despair because of the strict prescriptions that forced him to do what he saw as trivial research.

But, this does not alter the fact that with his paradox Barendregt described a problem of Dutch clinical psychology *as a whole*. There, in the course of time the dilemma of "humanity versus science," which as a matter of fact is expressed from time to time in many sections of international psychology, became increasingly obvious. Anyone who ascribes primary importance to "the" scientific standards – that is, the empirical-analytical ones – is quickly accused by practice-oriented colleagues of ignoring what really happens in everyday life and clinical practice, whereas anyone who emphasizes the provision of clinically relevant statements and theories is soon accused by his methodologically oriented colleagues of neglecting the rules of science. Although the neurotic paradox does not generally play a role in the *individual* lives and works of the other Dutch clinicians as much as it did in Barendregt's – some pay less attention to the methodological standards, others make fewer demands on the breadth or depth of their research projects – this subdiscipline as a whole also bears "two souls in one breast." As one of the participants in the discussion once noted ironically, Dutch clinical psychologists are divided between "the wicked who worship the Number" and the "good who acknowledge Man."[3] To a relatively large extent it is true of the professional field that it feels itself confronted with the choice between humanity and science, or in other words, between relevance and rigor. Barendregt therefore acts here as the *personification of the dilemma of the subdiscipline.*

In the previous chapters, the main question of this book – how the comparatively large methodological variability of the social sciences can be understood – was answered by investigating how in the history of Dutch psychology a particular group in a particular period came to endorse a particular methodological style. It was argued that in a methodological style not only a scientific identity but also a social identity is expressed. Methodological stability appeared when, and remained as long as, the psychologists concerned and their most important clientele agreed to this social identity, too.

The present example of Dutch clinical psychology confronts us with a more complicated case. Here we have on the one hand a gradual closure of the debates on what is to be called science and what is not, but on the other hand a remaining *uneasiness* expressed in the metaphor of the neurotic paradox. This chapter extensively describes the rather complex history of clinical psychology and the emergence of the paradox in the course of that history. Its central question is how Dutch clinical psychology fell into this dilemma of rigor versus relevance and why doubts about it did not go away. At the end of the chapter I conclude that the neurotic paradox is another example of a methodological issue that cannot be well understood without taking into account the social identity that is expressed in a methodological style.

Psychology at the service of "anthropological medicine":
The social identity of the first clinical psychologists
in the Netherlands

The history of clinical psychology in the Netherlands clearly illustrates that it is actually deceiving to speak of "the genesis of clinical psychology in country X or Y." This way of speaking suggests that clinical psychology was something with a given identity, *independent of* the concrete contexts in which it appeared or failed to appear, and that if the circumstances are favorable this "something" can establish itself somewhere. However, there are great differences between what was clinical psychology in a particular country in a particular period, and what was denoted as clinical psychology in another country or in another period.[4] Therefore it is difficult to define essential or fundamental characteristics of the discipline. In fact, what came into existence at various times and places was not one and the same thing, clinical psychology; rather, the *designation* clinical psychology became fashionable at various moments in diverse contexts for related but nevertheless divergent activities. For these reasons, it is not "the genesis of clinical psychology in the Netherlands" that will be described here, but the early use of the name "clinical psychology" there.[5]

The first professional association for clinical psychology was founded in the Netherlands in 1949. It is said that this "Society for Clinical Psychology" (*Gezelschap voor Klinische Psychologie*) took the name "clinical psychology" from the United States,[6] but it was certainly different in many respects from American clinical psychology. Immediately noticeable is that the foundation of a professional association for clinical psychologists took place much later than in the United States, as the American Association of Clinical Psychologists had already been founded in 1917. A second difference is that those who formed an association of clinical psychologists in the Netherlands were quite a different sort than their American counterparts. In the United States "clinical psychology" referred expressly not to the clinic as setting, but, according to Lightner Witmer, who already proposed the term in 1896, to a *method*. With the introduction of clinical psychology in the United States, a distinction was made between philosophical psychology on the one hand and experimental psychology on the other, in the sense that clinical psychology would concentrate on the study of individuals to promote their effective functioning. The first American clinical psychologists did some work with adults, but their main work was with children.[7]

In the Netherlands the criterion for membership in the Society for Clinical Psychology was in fact the *setting* in which the psychologist worked, and furthermore child psychologists were specifically excluded. Somebody could become a member of the society if he or she "devotes at least half of his [or her] activities to research of psychiatric or neurological patients including the psychosomatic, in close cooperation with a psychiatrist, and excluding specific children's practice."[8] Unlike their American namesakes, the first Dutch clinical psychologists were not united on the basis of methods that focused primarily on the study and education of individuals. For that matter, as shown in the previous chapters, with the latter they could not have distinguished themselves from the vast majority of the other Dutch psychologists at the time.

It is important to point out that collaboration with a psychiatrist was another necessary condition for membership. The addition of this criterion further distinguishes the early meaning of the term clinical psychology from that in the United States, where the very earliest clinical psychology was an initiative of psychologists.[9] However, in the Netherlands it was *doctors* who gave the first impetus to the establishment of a subspecialty called clinical psychology. These doctors did so for purposes they considered necessary and they themselves were also the first practitioners of the field. Moreover, the fact that they were a very specific type of doctor led to the most important distinguishing characteristic of clinical psychology in the Netherlands.

These were physicians who, rejecting "the dualism of Descartes," were opposed to a "purely materialistic vision of 'corporeality' [*de lichamelijkheid*]." Inspired by the early twentieth-century "anthropological medicine" associated with German physicians such as Binswanger, Von Gebsattel, Straus, and Von Weizsäcker, these Dutchmen developed under the same name a medical science in which "the person is more than just corporeality" and therefore not completely understandable from a synthesis of separated data.[10] In their quest to reunite body and soul, these doctors made the psyche as well as the body the object of medical care. This began at the Protestant Free University of Amsterdam, founded in 1880 "in protest at the materialization of science and against the use of the exact methods in the field of the humanities." Referring to the work of psychiatrists at the other Dutch universities who wanted to trace back psychic processes to bodily causes, at the Free University it was claimed that there must be a counterweight to the "unlimited influence of materialism in the medical faculties and in psychiatry at the national universities."[11] The Protestant doctors were of the belief that the soul must have its rightful place in physical medicine and psychiatry. On the basis of these ideas interest arose in the study of psychological processes.

In the Valerius Clinic, a psychiatric institution closely associated with the Free University, the physicians L. Bouman, H. C. Rümke, F. J. J. Buytendijk, and L. van der Horst attached great importance to psychology, which they chiefly understood to mean the study of the differences between individual people, and theories on the development of the individual. This psychology was mainly practiced in a medical context by the physicians *themselves*.[12] They taught psychology and they presented a part of their publications as psychology. At the end of 1922 they founded the Protestant Psychology Study Association (*Gereformeerde Psychologische Studievereeniging*) for "the practice of psychology, in both the philosophical and empirical sense, in the light of the Sacred Scriptures." In 1929 Van der Horst, who was professor of clinical psychiatry, also became professor of psychology at the Free University.

The role of the psyche in the genesis of illnesses also meant that the sick were in principle considered to be *accountable* for their diseases. From the perspective of these anthropological doctors "God honored man by making him always responsible." The psychiatrist/doctor therefore could not restrict himself to being a "physician in the narrow sense," because "whoever is seriously concerned as clinical psychiatrist with the treatment of the disturbed mental life, accepts the function of conscience as an extremely important impetus for healing and reintegration of the personality," and "Jesus also preached and healed the sick."[13] Apart from psychiatry and psychology the pastoral care of souls therefore had an important task in "strengthening or restoring the relationship of the sick person to God."[14]

On the basis of comparable ways of thinking, in the 1940s room was also made at the Catholic University of Nijmegen for psychological activities in a medico-clinical setting. As with the Protestant Free University and Valerius Clinic a new medical faculty of the Nijmegen Catholic University was founded to counterbalance the godless education offered elsewhere. The Catholic neurologist and psychiatrist J. J. G. Prick was asked to head the new clinic being built for neurology and psychiatry. Not long after his arrival in Nijmegen in 1940, Prick was also given responsibility for clinical psychology in the Faculty of Letters and Philosophy, under which psychology was then counted. This is how he became, as a neurologist, the first professor in the Netherlands to include "clinical psychology" in his teaching description and how, in 1942, the first training in clinical psychology in Europe got under way.[15]

Although Prick had studied at the neutral University of Amsterdam, he was certain nevertheless that medical science is or should be an "anthropological science." He had taken courses on this subject from Buytendijk, who converted to Catholicism in the 1930s. Prick and Buytendijk had become very good friends. Both of them worked on the development of a Catholic anthropological physiology and medicine, and they often worked together on this. In developing their ideas, they based themselves on the philosophy of the thirteenth-century philosopher Thomas Aquinas, which had been recommended by Pope Leo XIII in 1879 as general doctrine for Catholics. According to the ideas of Thomas Aquinas, the soul is the most important aspect of man, and the most important characteristics of the soul are reason and free will. The soul is also the immortal spirit, which reveals itself in an individual to the extent that the individual is rational and moral. The "corporeality of man" is, as it was expressed in a mixture of Thomistic and anthropological terminology, "animated corporeality" or "the incarnation of the spiritual life." From this standpoint Prick emphasized that the healthy person designs his own nervous system. To express this idea, instead of the term "nervous system," he preferred to use the self-invented phrase "the nervous systemness" (*zenuwstelselijkheid*) of man.[16]

Prick's special interest was in the role of psychic factors in the emergence of somatic illnesses. He classified various illnesses – ulcus duodeni, colitis ulcerosa, asthma bronchiale, multiple sclerosis, hypertension, rheumatism – as "psychosomatic," so he assumed that they were to a great extent the consequence of psychic problems. In Prick's words these were sicknesses that "are only apparently corporeal," and in fact "are an expression of a nonoptimal realization of existence, which is specific to being human." As the doctors at the Free University had suggested earlier, Prick, now inspired by Catholic philosophical anthropology, argued that patients are responsible for their sicknesses. Psychosomatic disorders were an expression of the *denial* of mind and will, which implies the denial of the soul: "Whoever actually expels the spiritual from man by acting not in accordance with the spiritual nature, will strip the body, which is for the sake of the soul, of its meaning and abandon it to decay."[17]

Prick's postwar plan to bring together anthropological medicine and anthropological psychology in an "Institute for Positive Human Studies" with a subsidy from the United States, could not be carried out because the funding failed to materialize.[18] However, in 1948 a Psychopathological-Neurological Institute got off the ground. The psychologist Van de Loo was appointed there as "clinical-psychological assistant." In 1952 Van de Loo got his doctorate under Prick's supervision with a thesis on "clinical psychology at the service of psychosomatic medicine,"[19] and in 1954 he became associate professor[20] of applied clinical psychology at Nijmegen University.

Elsewhere in the Netherlands in the late 1940s and early 50s the connection between medicine and psychology was also growing. Utrecht psychology students were given the opportunity to include psychiatry

Figure 9. J. J. G. Prick with a bishop (Mgr. Dr. J. M. J. A. Hanssen) and two priests on their way to the consecration of the Psychopathological-Neurological Institute, 19 June 1948 (photo from the Catholic Documentation Center, Nijmegen).

and anthropological medicine in their chosen subjects. One of the first to do so was Jan J. Dijkhuis. In 1947, on Buytendijk's recommendation, Dijkhuis was appointed to a Catholic psychiatric institution "for male nervous and mental patients."[21] He was wearing a white coat, was addressed as "doctor," and was given the task of assisting the institution's five physicians with psychodiagnostic research. This made Dijkhuis the first psychologist in a nonuniversity psychiatric clinic.[22] Soon, more followed[23] and at Dijkhuis's initiative some of these psychologists began to meet regularly to exchange experiences and tests. In 1949 this interaction resulted in the founding of the aforementioned Society for Clinical Psychology. In April 1954 the Society had eighteen members and nine candidates for membership.[24] A professional field called clinical psychology had been created in the Netherlands on the basis of the idea that the soul must play a role in medicine.[25] As far as I know, such a connection between a religiously inspired anthropological medicine and early clinical psychology is not mentioned anywhere else in the historiography of clinical psychology.[26]

It can be concluded that in its early years the social identity of Dutch clinical psychology was bound up with that of anthropological psychiatry. Clinical psychology's views on man and society were the views of anthropological medicine. Man was a creature with a soul, which implied that sickness and sin were closely connected. Obedience to God's laws was considered to be an important aspect of social integretion or reintegration.

Descriptive diagnostics: Methodological views of early clinical psychology

In the United States, soon after the American Association of Clinical Psychologists was founded, resistance to the activities of clinical psychologists arose from the psychiatric world.[27] In the Netherlands, where impetus for clinical psychology was actually given by psychiatrists themselves, such problems did not arise. However, as a consequence of the same situation the relationship between doctors and clinical psychologists was strongly hierarchical. The psychiatrists appointed psychologists as assistants in diagnosing and emphasized that they must "not in a sort of conceitedness play the psychiatrist."[28] They considered close cooperation with a psychiatrist necessary so that the psychologist continues to be aware of "the limits of his science."[29]

The latter point referred to the diagnostics that psychologists were not allowed to carry out on their own, but above all to giving treatments. The Dutch knew that in the meantime in the United States many clinical psychologists were working on the treatment of war neuroses and that for this reason the profession was undergoing a boom. The Utrecht psychologist Van Lennep had reported extensively and enthusiastically on this fact after making a trip to the United States in 1949. However, Dutch psychiatrists impressed the psychologists not to get involved with therapy. If psychologists were to give treatments they could "come in conflict with the law due to unqualified practice of medicine."[30] The psychologists themselves endorsed this hierarchical stance. For example, Van de Loo argued: "The psychologist who aspires to a fixed position in the field of the doctor must realize that he is not completely his own boss. This is inherent in his chosen status. The relationship of the clinical psychologist to the doctor (psychiatrist) cannot really be anything else but a hierarchically arranged one, in which the latter occupies the first place . . ." To fight against this would produce only "forced attempts at superiority and conflicts of authority."[31]

This form of psychology, which was so clearly at the service of antimaterialistic and antinatural scientific psychiatry, could not of course be oriented to the natural sciences itself. Another methodology would express its scientific and social identity. In his fierce pleas for a "new psychology" of the 1930s, the anthropological psychiatrist Van der Horst had contrasted his own approach to psychology in particular to that of Heymans's approach in Groningen. The new psychology, Van der Horst had stressed, emphasizes quality rather than the "measurable and quantifiable" and did not aim "to explain the personality," but "to understand it in its comprehensible coherence."[32] Until the 1960s this approach continued to guide the work of most Dutch clinical psychologists. Diagnostic methods that were originally psychiatric were used, including the ink blot test of the Swiss doctor Herman Rorschach, and patients were written about in psychiatric interpretative terminology.

Van de Loo's 1952 thesis offers a clear example. From case histories Van de Loo derived a general picture of "the hypertension patient" as someone having "protesting subordination" as a typical lifestyle. This attitude had its roots, he wrote, "in an early arisen and never resolved conflict between on the one hand strong aggression as a reaction of protest and disappointment to chronic deprivation of love and respect, and on the other hand simultaneous fear of authority and insecurity."[33] Rorschach reactions interpreted by Van de Loo confirmed this image. In the stories of patients with high blood pressure he identified not only more aggressive impulses than in those of a control group of healthy people, but the ink blot interpretations of the group with high blood pressure also pointed to an "unadjusted, unstable and egocentric emotional life, to a deficit in effect-stabilizing moments and to the presence of deviant wariness and ambitious perfectionism."[34]

As I described in the second chapter, the validation of tests began to gain ground in Dutch *selection psychology* of the 1950s, but in general clinical psychologists did not see much use for it. Some psychologists in that period wrote about *both* subdisciplines and then made a distinction in methodology. In a highly valued book on test psychology from 1952, for example, much attention was paid to "predictive tests" and to the statistical-technical sides of test construction, but hardly any function was ascribed to this in clinical psychology. There "naturally" those tests were considered important

> "which examine the affective, characterological aspects of the person. An exact and objective statement on the capacities or achievements is usually neither necessary nor desirable. More important is a qualitative analysis of personality, an image of the client in his stance with regard to the world, and with it a deeper insight into the motives which have led to the difficulties."

The "American tendency" also to quantify descriptive tests was called "irresponsible or meaningless."[35]

Predictive diagnostics: The clinical psychologist
as inspector of the psychiatrist

In the first postwar decade, research in the field of psychosomatics got also under way in the medical faculty of the University of Amsterdam. The internist Juda J. Groen, founder and leader of the Psychosomatic Working Group at this faculty, was not an anthropological thinker but a person inclined toward exactness. As he said in an interview many years later, in the fifties he did not have a high opinion of anthropological medicine at the Free University and in Nijmegen. What Groen, who had grown up in a non-Orthodox Jewish environment, could not believe in at all was the Christian researchers' central doctrine of psychosomatic illnesses: the existence of a separate soul. Neither his friendly contact with Buytendijk, whose courses in the 1930s he found inspiring "despite their vitalist bias," nor his friendship with Prick, could lead him to change his ideas on the subject. To him it was a mystery how his friend Prick could "as a man of science" believe "in religious dogmas like the immaculate conception of Mary and the real existence of a soul in every person."[36]

Groen had derived from American literature the "hypothesis of the psychosomatic specificity," which states that specific psychic problems lead to specific bodily problems in people who are susceptible to this through their personality structure, "just as when a tubercular bacillus penetrates an organism which is susceptible to that bacillus, specifically the sickness of tuberculosis appears." Groen sought explanations of psychosomatic problems exclusively in functional disorders of the endocrine-vegetative apparatus as a consequence of lengthy emotional tensions. In contrast to Prick he did not supplement his explanations with the "denial of the rational soul" by the patients involved.

Groen at least had more success in attracting U.S. funding than Prick did. In 1949 he had been able to found the Psychosomatic Working Group in the Wilhelmina Hospital with the help of the Rockefeller Foundation. Apart from Groen himself, a psychiatrist, J. Bastiaans, and a psychologist, S. J. Vles, had been attached to the working group. As elsewhere, it was the psychiatrist, assisted by the psychologist, who provided the diagnoses. Bastiaans subjected the patient to a study in the form of an interview that occupied three to ten hours. In this the patient told of his parental background, school years, and gratification in work and marriage. Not only the contents but also the form of the answers and the behavior during the interview situation were considered important for the diagnosis – besides, the assignment to associate freely provided information. Vles, the psychologist, administered the Rorschach to the patients. The reactions to the ink blots were quantified and the average deviation from the reactions of other randomly chosen subjects was calculated. Apart from supporting the psychiatric diagnoses, it was also intended that this would contribute to the testing of the hypothesis of psychosomatic specificity. To this end Vles sought the shared aspect in the Rorschach reactions of groups of patients with the same sickness. For example, he confirmed in this way the impression of psychiatrists and doctors that asthma patients are neurotic. Statistically significant differences between certain Rorschach interpretations of asthmatic and nonasthmatic youths had indicated that asthmatics have "a more infantile structure."[37]

It was quantitative, the hypothesis was confirmed, but nevertheless Groen was not much taken with Vles's approach. The way he used the Rorschach and interpreted its categories was, as Groen opined in 1987, "just like fortune-telling."[38] When Vles also wanted to introduce the Szondi Test in the working group, Groen's objections grew.

In this period De Groot gave the inaugural lecture mentioned at the beginning of this chapter, and Groen knew what kind of psychologist he really wanted to have. Johan Barendregt took Vles's place in the working group. Barendregt also used the Rorschach, but as a De Groot student he was also equipped with the empirical cycle. It had been instilled in him that carrying out research requires setting up hypotheses, deducing predictions from these, testing those predictions, evaluating the results, setting up new hypotheses, and so on. Barendregt introduced this methodology, developed in the context of personnel selection psychology, to the working group. He began to convince the other group members that his idea of testing the hypothesis of psychosomatic specificity on biographical case histories and psychoanalytic findings is much too subjective. Furthermore, he stated that the confirmations of the psychiatric interpretations already found experimentally in the group must be ascribed merely to the "trial-and-error method followed." In doing psychological research the group had constantly compared a number of Rorschach interpretations of the experimental group to those of the control group, without clear preceding hypotheses and precise predictions of research outcomes. He felt that a researcher should carefully specify beforehand what interpretations of the Rorschach plates he *expects,* given the clinical-psychiatric ideas. If he does not do so, he runs the risk of interpreting Rorschach results in the desired direction. Barendregt proposed going over to the De Groot method.

The idea that a certain psychosomatic illness correlates to a certain personality structure was chosen as the hypothesis to be tested. As De Groot had argued in assessment psychology (see Chapter 2), Barendregt emphasized that the "essence behind the observable" cannot be studied. Neither someone's soul nor his "personality structure" can be investigated, but only a number of *behaviors* that are supposed to be characteristic can. He decided to select the Rorschach as a means of eliciting the reactions (the "behaviors") against which the hypothesis could be tested. Reactions of asthma patients would be compared with reactions of stomach ulcer patients and healthy persons. Gradually the hypothesis was "translated into verifiable predictions,"[39] which ultimately resulted in seven specific predictions of Rorschach reactions for a group of asthma patients compared to that of both other groups. To be able to dismiss the null hypothesis – that there are no differences – would be seen as support for the hypothesis of psychosomatic specificity.

Whatever one's opinion of this project may be, setting up and carrying out such a study required specialized knowledge and skill in the field of methodology and statistics, which the doctors themselves did not have. This changed, inconspicuously but radically, something in the relationship between the two professional groups. Although it is true that Barendregt's conclusion supported the hypothesis of psychosomatic specificity, with the doctors of the working group being put in the right, that does not alter the fact that with these methods the capacity for judging the validity of diagnoses was removed from the physicians. Barendregt was no longer an assistant who completed medical judgments; he had worked his way up to be the psychologist who *controlled* the doctors. Besides, Barendregt introduced not only the new scientific identity of the clinical psychologist as a specialist in hypothesis testing. With this methodology he also endorsed the new social identity of a clinical psychology that instead of ascribing a soul to human beings, emphasizes the importance of their behavior only.

Testing psychoanalysis

The role reversal carried out by Barendregt was accepted in the Psychosomatic Working Group. Elsewhere, however, Barendregt's ways did not receive much approval. That could be seen in his second large project of the 1950s: a study commissioned by the Psychoanalytical Institute in Amsterdam (the PAI) into the results of psychoanalysis compared to those of other kinds of psychotherapy. Bastiaans, the psychiatrist of the Psychosomatic Working Group, had in the meantime become director of the PAI and he appointed Barendregt to carry out this study. In this way Barendregt – who at that time himself was in psychoanalytical learning therapy, too – got the chance to apply the new methodological insights to clinical practice with the PAI as well.

In the second half of the 1950s he further sharpened his methodological requirements. His chief source of inspiration was now the London Maudsley Clinic, where a student of Charles Spearman and Cyril Burt, the psychologist Hans J. Eysenck, had been doing experimental clinical psychological research since 1947.[40] Eysenck completely disregarded the psychologist's subservient role, which was also traditional in England, and subjected psychiatric insights to critical analyses. The psychiatric nosology, he argued, was derived from somatic medicine. Just as someone can have smallpox, cholera, or influenza, according to psychiatry a person could be schizophrenic, psychotic, or hysterical. Psychiatric categories refer to much less clearly distinguished phenomena than those from somatic pathology, though. With such an impure conceptual framework it was impossible to carry out research into the validity of diagnostic statements, for example, or the effects of various types of psychotherapy. In Eysenck's view therefore the psychiatric system should be replaced by a classification into several "fundamental dimensions of personality," on which each individual occupies a particular relative position. Just as every person is considered to be intelligent to a certain extent, according to Eysenck's approach every person is also to a certain degree neurotic, introverted, and so on. To the objection that psychology must take the uniqueness of the individual as its basic principle and that it is – using a pair of concepts from the philosopher Windelband – an "idiographic" and not a "nomothetic" science, Eysenck had stated that "It is quite undeniably true that Professor Windelband is absolutely unique. So is my old shoe." Uniqueness must not be regarded as some kind of mystical quality. The unique individual is "simply the point of intersection of a number of quantitative variables."[41]

The "fundamental dimensions of the personality" were tracked down by means of empirical research. Hypothetical constructions and theoretical concepts like instincts, motives, complexes, and attitudes had to be linked to reality via operationalization in terms of observable behaviors. To Eysenck's mind, only methods based on the study of "objective behavior" deserved to be called "scientific." This meant that those methods had to be extremely structured, should not demand extensive verbal answers, had to be capable of being quantitatively scored, and should not require "interpretation." Eysenck and his group constructed such tests, which they subjected to factor analytical research. This resulted in well-known "questionnaires" such as the "Maudsley Medical Questionnaire." The yes–no responses to questions like: "Are your feelings easily hurt?" "Are you rather shy?" and "Do you often feel just miserable?" determined someone's position on the neuroticism dimension.

Barendregt, who had made various working visits to the Maudsley

Clinic in the 1950s, was enthusiastic. Imitating Eysenck, he exchanged psychologists' traditional respectful tone toward doctors for a mixture of derision and severity:

> The statement by a psychotherapist that he has the feeling that psychotherapy has an effect somewhere . . . is of hardly any interest. This does not change when he can boast on wide experience nor even when he puts forward his conviction in fine statements. . . . An improvement in the situation of a patient after psychotherapy means nothing as long as it is not known how it would have been without that therapy. It is also meaningless that the therapist is convinced that the improvement is related to what was discussed during the treatment.[42]

Whereas in the first half of the 1950s the Rorschach still served as the criterion in testing the hypothesis of psychosomatic specificity, in later years Barendregt gradually rejected all projective techniques. In this respect, too, he distanced himself increasingly from psychiatric usages and customs. In his research into the effectiveness of psychoanalytic therapy, tests of *psychological* facture were used. For example, the Wechsler-Bellevue Intelligence Test was used to check the statement, which occurred frequently in psychiatric reports, that the intelligence of the patient is "depressed" by neuroticism. The Maudsley Medical Questionnaire was also used.

Following a "before and after study" of a group that had been in psychoanalysis, a control group that had had another psychotherapy, and a third group that had not been treated, Barendregt reported on the testing of his hypotheses. Each of the three groups had shown a sharp progress in subjective well-being. The hypothesis that those who were in therapy would progress more strongly than those who were not was not confirmed. The hypothesis that the intelligence of the two groups that had therapy would improve was not confirmed either, because the average of all three groups turned out to have risen. Furthermore the treated groups had not become less neurotic. In fact, a by no means negligible number of each group even scored higher on neuroticism.

The PAI psychiatrists fiercely impugned the research methodology. According to them the researchers had "with exaggerated bravura enthroned the empirical-analytical method."[43] Bastiaans and Barendregt responded that "a methodological basic training" was urgently desirable for psychiatrists and that "a work to appear shortly by Prof. De Groot [De Groot's *Methodology*] on the methodological foundations in psychology could lend a welcome impetus to the realization of this ideal."[44] That did not convince the doctors. No more subsidies were given to continue the research, and Bastiaans was also forbidden to publish the results.[45] Barendregt, who in the meantime had broken off his own psychoanalysis, now left the Psychoanalytical Institute as an employee, too. In his English-language collection of experiments, he included a short report of his research in the Psychoanalytical Institute.[46]

A foot in the door of the therapy room: Another social identity of clinical psychology

Even if Barendregt was the first clinical psychologist in the Netherlands to create a role for clinical psychology that was far more independent of psychiatry, his colleagues' reception of his work was just as unenthusiastic as that of the psychoanalysts. They discussed it under headings such as "A psychologist who wants to be so scientific that it is no longer scientific" and they rejected "the so-called scientific behavioral flavored psychology with its tendency to find statements on personality, unscientific 'waffle.' "[47] Barendregt's friend and colleague Nico H. Frijda defended him and accused his opponents of making "irrelevant contradictions" by "creating a contrast between the good who acknowledge Man and practice psychology, and the wicked who worship the Number and have nothing to do with psychology."[48] Barendregt himself wrote replies in which he curtly stated that the criticism was irrelevant.[49]

Meanwhile, the Society for Clinical Psychology had steadily grown. In 1956 it had been amalgamated into a clinical psychology section of the NIPP,[50] and in 1957 this section had seventy-nine members and candidate members.[51] Although it is true that the other clinical psychologists had not begun to gain an independent role for psychology along the route Barendregt had chosen, they made progress toward autonomy in the course of the 1950s by following another road. That was the road of mental health care. Remarkably enough, here once again it was doctors who in the first instance created the opportunity for psychologists. In order to explain that, I must take a short step back in history.

In the 1930s, inspired by the American Mental Health Movement, the Amsterdam psychiatrist K. H. Bouman had founded the Dutch Association for Mental Health, with the goal of "raising the public mental strength, maintaining the best psychic qualities of our community, the creation of conditions under which mental and moral values can develop, the striving to obtain personality types adapted as well as possible to their environment . . . for the benefit of individual and society."[52] What this was about, to put it briefly, was expanding the responsibility of medicine to people who were healthy (for the time being). Protestants and Catholics soon founded their own societies, with a comparable goal.

In the prewar years these societies had not yet really taken root, but after the Second World War they flourished by taking on the task of reconstructing the social life that had been disturbed by the war. Thus it could happen that medical terminology acquired a much broader area of application in the first postwar years than in the years before. Behaviors that were formerly condemned in the moral sense now became the subject of medical care. Divorce, for example, was called "not healthy," sexual behaviors considered undesirable required "healing," and it was thought that prostitutes should cool off in an institution. In short, behavior that formerly deviated from moral standards now deviated from medical norms: It all became a question of "hygiene."[53]

In Christian circles up until this time, overseeing the spiritual care of healthy people had been reserved for religious leaders. The innovation-minded doctors, however, defended the position that *psychic maturity* was a precondition for real faith and proper moral conduct. Punishment by the church would have to make way for treatment by medicine. For example, the physician Buytendijk emphasized in Catholic circles that a person should be able to choose faith and morality with full mental freedom and that for this he needed psychological guidance.[54] In Protestant circles it was also argued that mental health was a condition for real belief: Psychotherapeutic methods had to be used to "remove psychic disturbances which inhibit the effect of the Gospel."[55] Looked at in this way the task of the doctor preceded that of the priest.

Just as psychologists supported doctors in diagnostics they were also helpful to them now. They organized conferences on "Psychotherapy in the Light of Religion and Morality," they sat on working groups like the Commission on Pastoral and Mental Hygiene and the Pastoral Council, they wrote in journals on "counseling and spiritual care," and they published leaflets on "mental hygiene and Catholicism" or on "the healing power of love."

However, *while* psychologists were helping doctors to extend the area of problems that could be helped by science, they would soon begin to doubt the basic principle that help could be given only by doctors – the people who were being focused on might not be sinful, but neither were they really sick. These people could also be treated by psychologists. Now that mental health care had been separated from pastoral care, the next step would be to unlink therapeutic treatment somewhat more from medicine.

Along this road the Dutch clinical psychologists entered the field of psychotherapy. The first psychologist to "get a foot in the door in the competition on competence," as he put it himself years later, was H. R. Wijngaarden of the Free University. He did this by assimilating the concept of "conflictology," which also stemmed from the Movement for Mental Health. Conflictology was the study and treatment of "conflicts of a nonpathological nature." In 1951 Wijngaarden became associate professor of this field of study at the Free University, and in 1961, professor. With the name conflictology he consciously avoided the connotation with therapy. "That was safe with regard to the doctors," he later commented.[56]

The concept of conflictology never really became current. "Counseling," the "nondirective" therapy of the American psychologist Carl Rogers on the other hand, did. Van Lennep had already reported on it enthusiastically after his trip to the United States in 1949, in the 1950s Van de Loo had introduced the method in Nijmegen, and Dijkhuis started to use it in nonuniversity psychiatry.

Now clinical psychology had acquired a niche of its own. A new social identity for the discipline was embraced. The first people to be addressed were no longer the doctors but the clients of therapies; these clients were not considered to be sinful but in need of care and education. Social relations were now supposed to be regulated not by imparting awareness of moral guilt to people but by helping them to meet certain medical/psychological norms.

After the lines of demarcation between mental health care and religious care had been discussed and defined in Christian circles down to the last detail, and subsequently those between the doctor and the counseling psychologist in numerous publications and meetings, *behavioral therapy* soon followed as a specific psychological treatment technique. In the second half of the 1960s Barendregt was the great initiator of behavioral therapy in the Netherlands. Before this he still worked for a period on the development of psychodiagnostics.

Scientific research of the unique individual: The first appearance of the neurotic paradox

In 1962 Barendregt was appointed professor in personality theory at the University of Amsterdam. In his inaugural lecture he distanced himself from Eysenck's views.[57] On second thought the price of the London Maudsley Clinic methods had been too high. Barendregt now emphasized that a person's *uniqueness* was important in psychological research. However, that did not mean that he now acknowledged psychiatric interpretative methods. Barendregt was looking for methodological means that would preserve clinical psychology's own scientific identity while at the same time providing it with a social identity in which justice is done to individual uniqueness.

Once again, De Groot was followed, having earlier distanced himself from Eysenck's methods with the argument that "all data and predictions together" could still not produce a personal image of an individual. At the least, De Groot had claimed, the specific configuration and interrelation between the various data were unique, and it was precisely this unique structure that a clinical psychologist wants to get to know and understand.[58] De Groot's solution to this "problem of uniqueness" was in fact the same that he had given for assessment psychology in his inaugural speech in 1950, which was discussed in Chapter 2: formulate hypotheses on the individual involved, deduce predictions in terms of observable behaviors from it, and in this manner test the hypotheses. It was therefore testing without nomothetical principles; idiography *with* (probabilistic) statistics. Barendregt also argued that while maintaining the standards of scientific research, an idiographic method had to be used. He was also of the opinion, again following De Groot, that recognition of an individual's uniqueness made projective tests acceptable. He too gave them a place as a fruitful means of tracking down hypotheses about an individual.

In the Psychiatric Clinic of the Amsterdam Wilhelmina Hospital, Barendregt was given the opportunity to develop such psychodiagnostic means. Thus, as he later wrote, the decision was taken to study behavior that is usually called deviant, and so "to limit the field of personality theory [which was actually Barendregt's teaching task] to that of clinical psychology."[59] The new psychodiagnostic method used in Pavilion 3 of the Wilhelmina Hospital was given the name Experimental Individual Diagnostics (EID).[60]

What was new with regard to Barendregt's earlier way of working was that now the patients were no longer subjected to tests that determined their position on a dimension of neuroticism, or the like, which was supposed to be valid for all people. "Pavilion 3" also emphatically did not operate on the basis of any particular theory on people in general. It was thought a minitheory on each individual patient should be formulated, from which hypotheses would be derived on the causes of the patient's problematic behavior and those hypotheses would have to be tested with the help of existing means or new methods designed for the occasion. The factors tracked down could then offer psychiatrists points of application for treatment.

Barendregt appointed students as research assistants. In his house the working methods and results were discussed. An assistant at that time remembered him on these occasions sitting in an armchair, smoking heavily, and talking about the patients in a very inspired manner: "He was completely different then . . . he no longer cared about all those rules . . . then you saw all his imagination and his enormous ability to empathize . . . I have never seen such a great clinician."[61]

However, experimental individual diagnostics soon turned out to be not very productive. It was not possible to convert the imagination and empathy satisfactorily into testable hypotheses. A study sometimes took months. Each hypothesis required the designing of new testing methods. The diagnostic reports were as big as voluminous research reports that followed the pattern of experimental reports: formulation of the question, preliminary research with eventual theorizing, hypothesis 1, testing, results, conclusion; hypothesis 2, testing, results, conclusion; hypothesis *x*, testing, results, conclusion; and finally a summary, in which it was often stated that further research was needed. Reports of around twenty densely typed pages were not unusual. Over a period of five years, Barendregt and his assistants, helped by trainees, tested about two hundred hypotheses on around fifty patients. Having reached that point they abandoned the project. They came to the conclusion that "the method demanded too much from both the researchers and the patients."[62] "A single fool can invent more hypotheses than ten wise men can test," Barendregt wrote in retrospect: The approach "demanded more inventiveness and capacity for work than can be expected from anyone. Instead of hypotheses attuned to each patient it soon became the researcher's stereotypes stuck on each patient. Instead of operationalizations attuned to hypotheses it soon became hypotheses derived from the instruments simply present."[63]

The neurotic paradox had made its first appearance. Barendregt had not succeeded in convincing himself that it is possible to reconcile the study of people's "uniqueness" with methods derived from empirical-analytical assessment psychology. "To reduce dissonance," he wrote later,

> the causes of the failure were sought in the imperfect psychiatry and in the dependent position of psychology in a psychiatric environment. It was decided to work more autonomously by fixing the object of research ourselves and thus limiting it, and by attending to the therapy ourselves. In this way the psychiatric clinic of the Wilhelmina Hospital was to a great extent abandoned as a field of work.[64]

Barendregt then got the idea of the "phobias project." The research and teaching program of this university department was set up around the phenomenon of the phobia, described as "the showing of anxiety and avoidance behavior in situations not objectively dangerous." Once again the Maudsley Clinic was a source of inspiration. In experimental research into the explanation of deviant behavior, these London psychologists had applied principles from psychological learning theory. This led them to the idea that these principles could also be used to induce behavioral change. They had therefore developed a therapy based on psychology that they promoted as an alternative to "labor-intensive and furthermore failing" psychoanalysis, which, to indicate the contrast with the psychiatric "psycho" therapy, was called "behavioral" therapy.[65]

In the *International Journal of Psychiatry* Barendregt criticized Eysenck's "black and white representation" of the advantages of behavioral therapy, and he wondered out loud whether behavior therapy uses "much more of modern learning theory than the not so modern blows for unlearning and sweets for learning." Besides, he was of the opinion that "one often starts on a patient with a clear principle of learning theory only to wind up with a vague Freudian transference."[66] Nevertheless, his phobias project would be based primarily on behavior therapy techniques. Students were pouring in by now in large numbers.[67] Unlike learning to psychoanalyze, learning to apply behavioral therapy techniques did not take years of training. With the lighter cases the institute would treat, students would be able to work with these techniques after a comparatively short training course. There were methodological advantages, too, Barendregt argued, because the principles of learning theory are more explicitly formulated than those of psychoanalysis, and the effects of behavior therapy could more easily be studied scientifically.

The dissemination of the neurotic paradox

When Barendregt devoted himself in the early 1960s to experimental individual diagnostics, and especially when he practiced therapy outside the medical setting in the late 60s, his independent stance was no longer an exception in Dutch clinical psychology. The criteria of empirical-analytical methodology were increasingly accepted by the psychological community as the rules of science in general. For more clinical psychologists than just Barendregt alone, this implied an escape from the patronage of psychiatry. At the same time the institutionalization of "nonmedical psychotherapy" and the development of psychological-therapeutic techniques offered extra possibilities for the independence of psychiatry. However, these two "escape routes" together would lead many more clinical psychologists to fall into the dilemma of scientific justification versus practical relevance. The neurotic paradox began to manifest itself in the profession of clinical psychology as a whole.

When looking for methods and techniques, clinical psychologists attached to the universities in particular were no longer following the mores of the psychiatrists but rather those of their fellow psychologists. For instance, the Maudsley method of Experimental Individual Diagnostics was also proposed at the Institute for Clinical and Industrial Psychology in Utrecht. In the bulletin *Announcements of the ICIP (Mededelingen van het ICIP)* the Maudsley tone made its appearance. Instead of being a "diagnostic extension"[68] of the psychiatrist, the psychologist had to acquire his own task and function by applying his scientific skills.[69] For this the psychiatrist would have to be trained to come up with clearly formulated questions. For example, when a psychiatrist asks: "Will you give the Rorschach, so that you can help me to differentiate between schizophrenia and neurotic obsession?" the psychologist should not give the Rorschach. He should explain to the psychiatrist that the psychiatrist has to think about his question again:

> We know after all that these terms differ in content from psychiatrist to psychiatrist. Furthermore, it could be the case that the problem presented by the psychiatrist is not at all a problem. There is no known scientific reason why a patient cannot be both obsessive and schizophrenic, as these terms refer to extreme positions in personality dimensions, which do not correlate.

The psychiatrist must "be helped to make himself more explicit," the problem had to be "stripped of obscure terminology," the goal of the research must be clarified, and the psychiatrist must "be informed of scientifically known facts."[70]

As for Barendregt in the fifties, the empirical-analytical rules now also bestowed autonomy on the clinical psychologists. However, at the same time, diagnosing became an increasingly less popular task. Psychologists were of the opinion that a psychiatrist only did something with the data given if these data were congruent with the opinion that the psychiatrist already had: "He reads once again in black and white that it has been confirmed with complex and above all scientific methods what he, the clinician, had already concluded long ago."[71]

Psychologists wanted "new challenges" in their work. Furthermore, as independent therapists they would be in a better position to negotiate the abolition of the large difference in salaries between doctors and psychologists.[72] The rapid growth of the welfare state, in which the government organized professional help on a larger scale, also meant that psychotherapy became a facility available to increasingly large groups of people.[73] This gave clinical psychologists the possibility of focusing more and more on the expansion of nonmedical psychotherapy. In a survey among clinical psychologists in 1961, 52 percent of the respondents mentioned counseling or individual therapy as part of their tasks, and 30

Figure 10. Cartoon published in the Journal of the Dutch Association of Psychologists (*De Psycholoog,* June 1969, p. 316). Arend Van Dam, Studio 21, Nijenrode.

percent cited group therapy.[74] In this fashion diagnostics – formerly *the* means for the psychologist to profile himself in the medical-psychiatric world – gradually became less attractive when clinical psychology began to develop its own insights and skills in the therapeutic domain.

However, these were not the only reasons for dismissing diagnostics. In the context of their *own* therapeutic work clinical psychologists also saw less and less benefit in administering tests. They could not reconcile the role of a counselor with that of a forecaster of behavior. Instead of helping to substantiate or refute predictions as a good counselor would do, the therapist who wanted to diagnose in a scientific manner would have to wait and see if his predictions came true. In addition, scientific testing of a diagnosis demands that it be hidden from the client so that the client himself would not change his behavior before the testing process is complete, but according to the doctrines of counseling, withholding ideas from the client would damage the client's trust or otherwise interfere with the therapeutic goals of his individuality and independence.

Jos. H. Dijkhuis, who in 1963 was appointed lecturer, and in 1965 full professor in "nonmedical psychotherapy," endorsed a predictive approach in psychodiagnostics, but at the same time stressed the point that as a *treating* psychologist one should not practice scientific diagnostics. Then one has to fall back on "what the clinical approach was originally": "The art of choosing adequate and relevant verbal expression in dealing with people, or characterizing with adequate formulations people's rela-tions."[75] In the context of treatment "it is not the objective 'truth' . . . which is important but the relevance for the client himself. The client has to be able to use the data, the verbalizations must appeal to him."[76] Insofar as it was necessary to make a diagnosis, more clinical psycholo-gists argued at the end of the 1960s that it must be a diagnosis that the client makes, together with the therapist, about himself. That does not happen in a diagnostic phase preceding the therapy, but during the self-knowledge phase promoting therapeutic process.

In this way empirical-analytical methodology soon came under discus-sion again. This was no longer because diagnostic reports of the essential aspects of a person were preferred and thus a *different idea of science* was defended. Science was now associated with the rules of empirical-analytical methodology. The psychologists' problem was that the princi-ples of scientific diagnostics did not coincide with those of good therapy. So at the end of the 1960s the "dilemma of humanity versus science" appeared in clinical psychology as a whole, as the field, in trying to achieve an independent status, simultaneously converted to empirical-analytical methodology, *and* applied itself to giving therapy according to nondirective principles. Again, predictive empirical-analytical methodol-ogy collided with the social identity of clinical psychology.

Barendregt and "the phobias project"

In a survey of NIP (Dutch Association of Psychologists) members in 1970, 65 percent of clinical psychologists stated that they carried out psychotherapy and 35 percent of them indicated that they would like to do it.[77] In a 1972 study into the functions performed by psychologists, it even emerged that in psychology as a whole, psychotherapy appeared to win over psychodiagnostics: In 13.8 percent of the cases psychotherapy was named as the most important task and in 8.9 percent of cases diag-nostics was chosen (as opposed to 3.3 percent and 19.8 percent, respec-tively, in the same report in 1954). In 1972 psychotherapy even scored the highest of all fifteen possible functions of psychologists that the pollsters had summarized.[78]

"It seems to be a very satisfying activity," sneered Barendregt:

to make good money from it and then also be called engaged: *ein guter Mensch sein, ja wer wär's nicht gern*. Others apart from clinical psychologists have also in the meantime beheld this para-dise. Child psychologists already kept apace of clinical psycholo-gists. But now the social psychologists are also becoming thera-pists; and the industrial psychologists are becoming therapists. The experimental psychologists are now dealing with functional disorders; they therefore possibly also become therapists. All psychologists are becoming therapists. . . . But whether this makes them brothers is not certain.[79]

In order to raise the question about the effect of all those psychotherapies above "the question of whether bloodletting or mudbaths are better for a fever," Barendregt was of the opinion that not all psychologists should want to be therapists. At least some of them will have to concern themselves with research. He mainly wrote methodological commentaries on this, deriving his arguments from experiences and research results from the phobias project. Concentration on only the phobias and on behavioral therapy appeared to be not enough to dispel the phantom of rigor versus relevance.

Once again, in Barendregt's view, the quest for methodological clarity had led to too much reduction of reality. He wrote about this in several articles.[80] He pointed out that in the phobias project, they soon came to the conclusion that a phobia is a far less unambiguous phenomenon than it first appears to be. The description of a phobia that is useful for research, as "fear in and avoidance of a situation not objectively dangerous," was not sufficient in clinical practice. Barendregt gave examples. People who avoided situations not because they were anxious, but because they became angry came forward. For instance, there was someone who flew irresistibly into a rage when he heard somebody sighing or breathing heavily. There were also people who were not able to wait for a short while in any situation without becoming furious.

It was very urgent, in Barendregt's opinion, that something be done about the one-sided orientation to behavioral therapy in the phobias project. It turned out to be necessary to go on looking for underlying causes: not only did people avoid extremely diverse situations, but the avoidance of the same or comparable situations turned out to have very different possible backgrounds. Street anxiety can be caused by undiagnosed bodily disability that leads to dizziness on the street, but also through inaccurately localized tension that in fact arises from a combination of factors such as bad housing, a bad marriage, and money worries. Fear of trains can have its origins in a train journey to a feared destination, but also in all kinds of other factors that realistically have nothing to do with train journeys. Just as the symptom of fever can have very different causes, one agoraphobic is not the same as the other. Phobics turned out to have their own souls. In-depth diagnosis was needed: not only the factors that produce an unpleasant experience must be traced, but also the factors that led the patients mistakenly to localize the source of their misery, and apart from this the factors that make the patients stick to their mistakes.

Barendregt also rejected the idea that all the different existing therapies actually boiled down to the same essence and that therefore it could be fruitful to look for a translation of one therapy into that of another. The various therapies, in his opinion, were not derivatives of a fundamentally similar medicine, but different remedies for different complaints. Just as the same phobias did not necessarily refer to the same sickness, not all therapies were suitable for all phobias, and therefore there was not one particular therapy that was automatically the most suitable means for healing every phobia:

> Judging by their often unbearable in-group language, with which they pronounce condescendingly on other therapies, it must be assumed that the representatives of various schools think their methods best for practically everyone. A preliminary diagnosis is therefore not necessary. Many therapists consider a diagnostic study not only superfluous but even disturbing for the treatment. After all they already know that the other patient too will once again confirm their ideas. They have more important matters to think about: the dissemination of a message with which humanity can be saved. In this way therapies are oriented more to the therapists than the patient.[81]

In order to test Barendregt's ideas it would be necessary to work with more types of therapy than just behavioral therapy in the phobias project and attention would have to be paid again to diagnostics. In paraphrasing Eysenck's "what is true in Freud is not new and what is new in Freud is not true," Barendregt expressed once more the dilemma of rigor versus relevance: "What is true in Eysenck is not interesting and what is interesting in Eysenck is not true."[82] After all it was the task of psychology to make statements on "the human soul," "the principle responsible for the coherence and continuity in the acts of a person, which nowadays bears a scientific name: the structure of human personality."[83] For every phobic there would be his own personally tailored diagnosis and therapy. That would be the ideal situation. But this was not feasible. Referring to the failure of the experimental individual diagnosis in which in principle an infinite number of categories was available, he argued that it would be good to limit the possibilities now.

In this way Barendregt arrived at the proposal of a diagnosis according to the "forms of organization" central to various existing therapies. Somatic therapy seeks the causes of psychic problems in bodily aspects and latches onto this in treatment. The psychoanalyst seeks it in blocked goals: solving incompatible goals is central to its therapy. Behind ego-therapy is the idea that (role) behavior can be derived from self-image and position: Acceptance or change of this and the learning of role competence is prescribed. Relationship therapy tries to find it in the individual's involvement with his immediate social surroundings and tries to improve undesirable forms of relationship. Finally, behavioral therapy sees the habit as the most important form of organization and wants to teach new habits and abolish the old.

That is what the most important therapies have to offer and that is what diagnosis should try to achieve. A diagnostician is supposed to find out why someone ascribes the source of his suffering to the wrong factors. He must also have a physical examination carried out. He must ask himself whether it is mainly a case of inner conflicts, or too low self-esteem, or discontent with a social position. He must also examine the relationship with parents and partners and must ask himself to what extent the phobia has become an autonomous habit. Usually there are complaints in more of these areas because one evokes the other. However, according to Barendregt it should be a matter of finding out which of the forms of organization is most "in disarray." This, then, is the first point of attack for therapy.

The program demanded that the forms of organization of the various therapies be made properly explicit, that the characteristics of people that make a form of organization vulnerable be tracked down, and that good operationalizations be found for that. Only when all that preliminary work had been carried out could the effect of the implemented therapies be examined.

In his articles Barendregt reported on the first steps that were taken in this program. However, the reports did not go much beyond the first steps. Many problems arose. One of the most important was that the "unbearable in-group language" he had ironized before made way for "Ecunemicism in psychotherapeutic Babel."[84] In the course of the 1970s the boundaries between the various types of psychotherapy became vague. Psychotherapists called themselves "eclectic," by which they meant that they were not committed just to one sort of therapy, but used a mixture of the various models: here elements from psychoanalysis, there a piece of behavioral therapy, nondirective conversations if required, and so on. In this way, the practice of psychotherapy made pure testing according to Barendregt's plans impossible.

In the phobias project, to which in the second half of the 1970s at least forty-five researchers had already been appointed, the mutual contact and attunement between researchers began to diminish. Each had his or her own cares and tasks and the idea of a large-scale communal project did not amount to much. Researchers and therapists especially drifted apart.[85] Barendregt interpreted differing opinions from his own staff as attacks and wrote sharp personal reactions. There were few project members at this time with whom he had good contacts. He withdrew into his office and took less and less notice of what was going on in the department.

In this period he was working on a little book entitled *Characters Of and After Theophrastus (Karakters Van en Naar Theophrastus)*. In this work he translated Theophrastus's famous characters into parodies of the mores of *psychologists*. The hypocrite, the dandy, the eager beaver, the prig, the boaster, and the arrogant all get their comeuppance in it. The piece on the superstitious begins with the thesis: "Superstition is the veneration of the God of Methodology" and continues with "the superstitious is capable of finding fruitless research perfect; 'the fear of method is the beginning of wisdom,' he murmurs the entire day, and to cleanse himself from the enchantment by psychoanalysis he daily practices meaningless syllables."[86]

In the late 1970s the tone of Barendregt's articles changed. His pugnacity drenched with cynism made way for despondency coated by irony. In his article that defines the neurotic paradox (quoted at the beginning of this chapter), he also considered that "methodology consists only of parables": "Such a parable actually only means that you cannot learn from anyone how it should be done in psychology, that you have to find out for yourself; and you knew that already. Because every day you wear yourself out with very different matters than those which a methodologist can talk about so well." But: "Another time you think that you would not want to be without a parable like that of the methodologist. Because so much is said about the soul and often the people say anything that comes into their heads: in order to be agreeable to others for example, or disagreeable."[87] With the same hesitation he now distanced himself from "the commandment: thou shalt predict":

> Apart from being just, the commandment is also strict. Certainly for psychologists if they do not want to confine themselves to trivialities. Or if they are more interested in the activities of existing people than in the difference between two figures which stand for averages of strongly overlapping groups with a large distribution. Sometimes psychologists think they really do see connections. But there is always a disturbing factor which makes their predictions ridiculous. It is enough to make you disheartened. And that cannot be the intention of methodology.[88]

The flowering and stagnation of psychotherapy

At the time that Barendregt was setting up the phobias project and then making fun of the fact that all psychologists wanted to be therapists, most other clinical psychologists were busy with activities leading to the social recognition of their new professional role. They had meetings about ways of achieving professional accreditation, about the specific postgraduate training requirements that had to be made of a real clinical psychologist, about the question of whether or not they should remain members of the NIP or as therapists should distinguish themselves from the other psychologists and unite under the Dutch Association for Psychotherapy (*Nederlandse Vereniging voor Psychotherapie* or NVP).[89] They lobbied the government to lay down various points in law. In short, they developed all the activities that were necessary to profit in the development of the welfare state, in which the care of other people was increasingly becoming a state matter.[90] Through negotiation with doctors, for example, it was achieved that the payments, fixed at the government level, were equal for all psychotherapists, irrespective of their training. That made the profession of psychotherapist – already the best paid kind of psychologist – much more lucrative than that of an "ordinary psychologist."

In this fashion, in the first half of the 1970s the creation of an individual scientific identity was not the first priority. Diagnostic and accompanying methodological questions hardly appealed to the imagination. Even more strongly than before the prevailing social ideals prescribed a nonobjectifying interaction with people: The "humanistic psychology" begun by the American Abraham Maslow, with its image of man as motivated self-actualizing being, was disseminated. And the Marxist-inspired "critical psychology," which gained a following in the 1970s, made "objectifying research" unpopular from a more political version of such philosophical-anthropological ideas. Diagnosis had practically disappeared from the teaching program of clinical psychology departments. It is true that some did resent alternative forms of "diagnosis" that take place during a treatment process,[91] but apart from Barendregt only a few others at this time argued in favor of clearly separating diagnosis from therapy, according to the established scientific criteria.[92]

In the second half of the 1970s, this situation changed. At that time the Netherlands was devoting the highest percentage of its national income, compared to all countries in the world, on social expenditure.[93] With the advent of an economic recession the idea of devising a system whereby the state is responsible for the welfare of each individual came under discussion. The welfare state was said to release individuals too much from their responsibilities. The state, aided and abetted by professional meddlers, was reproached for getting too closely involved in private matters. Psychotherapy in particular began to lose much of its obvious goodwill. Questions were asked about its efficiency and economizing measures were taken.[94] "The fall of the psychotherapeutic business" was already forecast.[95] At the end of the 1970s the measure that required equal payment to medically and nonmedically trained therapists was being repealed again. In the course of three years the honorarium of nonmedically trained therapists was reduced by about 6 percent per annum.

Thus growth came to an end. This recession began while clinical psychology was producing increasing numbers of graduates; unemployment began to appear among clinical psychologists. For example, whereas the 1972–3 graduates from the phobias project all found work, 18.2 percent of the batch from 1977–8 were still without a job by 1980. At the beginning of the 1980s unemployment among Dutch psychologists was 25 percent, 40 percent of whom were clinical psychologists.[96]

And so it happened that various authors explicitly referred once more to the employment possibilities offered by diagnostics.[97] Warnings were issued that other professions were filling up the "gap in the market"[98] and psychologists pointed out to each other that it is above all because of their "methodical competence" that they have more to offer than related professional groups.[99]

Professional journals once again published methodological treatises. Some of these explicitly discussed the neurotic paradox. Much attention was given to "why the effect studies are often such an abomination to the methodological eye."[100] And the answer, given in diverse formulations, was that great difficulties are encountered in the attempt at methodological precision in researching relevant questions.[101] In 1981 the conference theme of the Dutch Association of Psychologists was also "the tension between scientific psychology and the psychotherapeutic practice." Some papers presented at the conference had been published beforehand. It is striking that each of these texts emphasized that "the traditional effect-research" is badly tailored to psychotherapeutic practice.[102]

Various proposals were made to escape the neurotic paradox. An advisory group set up after the conference argued that clinical psychologists would have to retain the professional field of therapy-research along alternative routes. They could carry out research on the cost saving of psychotherapy in somatic medicine, on values that were communicated in psychotherapeutic help, or on the failures and near-failures of therapists, and so on. The professional association would have to promote the financing of these research projects by university "science committees" and by subsidy givers, who were still oriented only to the "antiquated model" of effect-research.[103] Some now spoke dismissively of "the preoccupation with reliability of measurements and the keeping under control of variables" and argued that the criteria laid down for proper psychology had to be put into perspective.[104] A lance was also broken on behalf of the old projective methods.[105]

With such proposals the choice was made to broaden the criteria for science, that is, to look for a less strict scientific identity. But at the same time solutions were being sought at the other side of the dilemma. While retaining empirical-analytical criteria of science, some devoted themselves to precisely defined and relatively limited questions on behavior determinants, diagnostic methods, or the effect of a therapeutic measure.[106] By publishing with Anglo-American publishers and in English-language journals, these clinicians achieved a relatively large research output. In the publications-directed climate at the Dutch universities of the 1980s, they thus acquired most tenured academic jobs in clinical psychology.[107]

"From Eysenck to Musil"

From the late 1970s on Barendregt paid little attention anymore to the subject of science and psychotherapy. He was working on a new project: the design of a theory. He was still wondering what actually animated the patients whom he saw around him. He formulated an explanation of the source and core of phobic anxieties and he did that in a manner more philosophical-literary than methodical.

The clients' accounts at the intakes had given him an idea. It had struck him that not only the diversity of the anxiety-inducing situations, objects, and states is extremely great, but also that many subjects described their fear in vague terms like "something ominous," "that other," or "it," but not in the more concrete "being afraid of something."[108] The experiences described by the phobics, according to Barendregt, are the same as the phenomena known in psychopathology as "acute depersonalization experiences" and in philosophy and literature as "existential anxiety" or "alienation," "nothingness, the void, the abyss, the chasm of the soul, the Great (Rilke)." Moreover, it is remarkable, he thought, that the phobics are not really depersonalized, as happens in a delusion or psychosis. They do not say that the houses *are* of cardboard, but they say it was as if the houses were of cardboard; they do not claim that they *are* a haywire computer or that their face *has* disappeared, but say they feel as if this were so.

Figure 11. Portrait of J. T. Barendregt by Anne Hees (*Het Schaakbulletin*, 167, October 1988, p. 11).

Barendregt's theory was that phobics escape from the threat of real psychosis by seeking something to hold onto in an anxiety, which is to be preferred to the real void. The phobic avoidance behavior that subsequently appears is a way of dealing with the anxiety. This must be rationalized and canalized. The phobic preserves himself from losing, "with apologies to Musil," not his character traits but his person, just as a melted snowflake still consists of water but has lost its structure.

Barendregt was frustrated that his theory was not really verifiable. Nevertheless, he was now able to approach the methodological criteria with such freedom that he did publish his ideas, even if he distinctly referred to their incompleteness.[109] When he died in 1982, fragments of his theory were ready for publication in a more developed version. His son, the mathematician Henk Barendregt, assembled these fragments into a whole paper. In the collection of essays known as *The Soul Market (De Zielenmarkt)*, which had been in preparation long before Barendregt's death but was published posthumously, this paper was included as the last chapter.[110]

"So the path led from what is sharply definable to something for which words and concepts must be sought and borrowed; the path from Eysenck to Musil. So the way was chosen from what is, or appeared to be, verifiable, to what is important . . ." wrote his friend and colleague Nico Frijda in an obituary.[111]

The genesis of an occupational neurosis

"The road from what is verifiable to what is important." To the last, verifiability and importance remained incompatible for Barendregt. The combination continued to be "a romantic image." Only by stopping to strive for rigor as well as relevance could falling into the neurotic paradox be avoided. Barendregt finally did that by ignoring – what he saw as – the criteria of science. This proposition was chosen by other clinical psychologists at the end of the 1970s, too. There were also clinical psychologists who did not so much ignore the prevailing criteria of science as those of relevance. The appreciation of these "escape strategies" is here left to the clinical psychologists themselves, who to this day accuse each other, and sometimes themselves, of either irrelevance or lack of methodological rigor. To the present historical analysis another problem was central: that is, how in the course of its history did clinical psychology end up in the dilemma described?

To anyone who has experienced the neurotic paradox personally and thus looks at the problem *from the inside,* it is inherent to clinical psychology: The central questions of this discipline are so complex that they cannot be handled with scientific methods. The human soul, the first subject of clinical psychology, was not to be grasped with scientific means nor could it be pinned down by science in its secular form of "the structure of personality." Insofar as the rules of science are followed, questions about what really motivates people are inevitably transformed by methodological treatment into residues of the real problems, and the resulting answers are accordingly unusable.

Such an explanation makes eternal and universal verities of both a particular view of science and particular "central" research questions. *From the outside* the dilemma looks different. In short, in this chapter the genesis of the neurotic paradox was sketched as the unintended and unforeseen outcome of the two historical lines along which clinical psychology liberated itself from the patronage of medicine.

The clinical psychologists fought for their emancipation from psychiatrists by assuming an autonomous social identity in the context of mental health care. They became counselors and subscribed to a vision of society in which people were governed by learning to meet certain psychological norms. Individual freedom and responsibility were important premises. Besides, the clinical psychologists became researchers who wanted to understand the uniqueness of the person, the inner structure of personality, and the like.

However, clinical psychologists *also* fought for independence by borrowing tried-and-tested methodological rules from elsewhere in psychology – mostly from empirical-analytical assessment psychology. As described in Chapter 2, along with the exchange of the empathizing *geisteswissenschaft* methods for a calculating empirical-analytical approach, the model of the counselor had been exchanged for that of the detached middleman in manpower. In the era of new postwar labor relations, the question of who an applicant really "is" was replaced by the question of what work performance he or she could deliver. The notion that with predictive methodology "the most essential aspects of a person's being in the world" could no longer be an object of study was not at all deplored.

This sheds a different light on the problem of why no stability arose in clinical psychology, or why this subdiscipline was continuously plagued by doubts. In their battle for autonomy the Dutch clinical psychologists assumed the scientific identity of predictive empirical-analytical methodology, but they could not reconcile themselves to the *social* identity implied in this methodological style. In clinical psychology, one might say, the idea that "the most essential aspects of a person's being in the world" was no longer important *was* seen by many as a loss.

In the previous chapters an important condition for stability appeared to be that psychologists and their clients could also subscribe to the social identity expressed in a methodological style. This was precisely the condition that clinical psychology failed to meet. In this light, the dilemma of rigor versus relevance is not the inevitable outcome of an essential and universal incompatibility of science and humanity. It is an *incompatibility of social identities,* an incompatibility that emerged historically in the laborious process of clinical psychology's emancipation from psychiatry. The neurotic paradox is another example of a methodological issue that cannot be properly understood without taking into account the social identity that is expressed in a methodological style.

5

Predictions

In order to investigate what the three previous historic case studies have produced up to this point, it is illuminating to review the line of reasoning from the beginning. Chapter 1 discussed the claim that methodological changes in the history of Dutch psychology can be ascribed to a steady growth of insight into the proper rules of science, that is, that slowly but surely the Dutch also came to realize how psychology must be practiced if it is to be done in a scientifically correct manner.

It was also argued that historical work that shows such a process of growth generally does so by declaring, implicitly or explicitly, the contemporary state of affairs as its criterion. Those aspects of the past that deviate from present-day criteria are rejected as a murky detour and extra attention is given to that which "already" met the criteria of the present. In finalistic accounts arising in this way, an objective lineal growth of insight from the past to the present is suggested, whereas in fact the line is drawn by the historian her- or himself from the present to the past.

This led to the question of whether or not the image of cognitive progression could also be maintained in a historical study where the contemporary state of affairs does not function as a criterion. Put another way, the question was whether, in a nonfinalist approach, the image of the slow but steady discovery of what science is would survive. This is not excluded as a matter of course. In principle, in a nonfinalistic approach, too, systems of now-antiquated rules might turn out to have been inadequate and a new methodological style could prove to have been more suitable.

However, this conclusion would mean that there were periods in the history of Dutch psychology when academically trained and experienced practitioners of the discipline, who furthermore were aware of better methods via the professional literature at home and abroad, persisted for decades in making errors that were against their own interests. Their clientele would also have accepted an unusable product all through that time. Such an unlikely conclusion can be convincing only when it has turned out to be inevitable. Therefore it was decided first to try and make plausible what could be called the *inherent* rationality of ideas, which now seem strange. At the end of the first chapter the alternative hypothesis was developed: that Dutch psychology in the course of its history affiliated itself to different social groups with different ideas on social relationships and that this sociocultural pluralism was expressed in its variable methodology. Thus, in short the claim to be researched became that methodological styles express a social identity.

In the three succeeding historical chapters this image remained intact. In each case it appeared that agreement on the social identity of the profession was a precondition for consensus on the methodological level. In Chapter 2 the variegated pieces of the historical jigsaw fitted into a picture of affairs in which the *geisteswissenschaft* assessment psychology was understandable or even rational in the light of the social relations it supported, and in the third chapter a similar image could be built up with regard to the Utrecht phenomenological psychology that appeared to be an academic translation of the ideals of personalistic socialism. The social identity turned out to be important in Chapter 4, too. There, existing methodological dissatisfaction in clinical psychology could be ascribed to the underlying lack of consensus on the social identity of this subdiscipline.

Can it be now concluded from all this that the idea of gradual awareness of only one correct methodological style must be abandoned? Can it be also concluded that the epithet "scientific" cannot be withheld from methodological styles such as phenomenology and the *geisteswissenschaft* psychology, now that they appeared to be convincing seen from their own context? Even if one agrees with the plausibility of my historical reconstructions, the answer to *these* questions is not yet necessarily affirmative.

The objection could be raised that it is still possible to say that Dutch psychology has gradually become scientific in the sense that it has finally joined the image of science that just happens to be the *internationally current* one. It was only with the implementation of hypothesis-testing methodology that Dutch psychology really became a science because its practitioners then pledged themselves to follow the rules that happen to apply – speaking merely empirically – for science all over the world. That the appearance of other constellations of rules in the course of Dutch psychology is understandable does not mean that they are also scientific.

If the historical developments from the preceding chapters are viewed from this perspective, then it could be said that the tendency to a mature science had already commenced with Heymans at the end of the nineteenth century. This would then come almost to an end in the heydays of the *geisteswissenschaft* and phenomenological approaches, which were able to emerge under particular social circumstances. The right approach would be further developed again only at a later stage. Even though the social circumstances under which current Dutch psychology established itself could be of a local nature, it is nevertheless a textbook example of the *universal* nature of scientific rationality. Looked at in this way the history described up to now shows at best that certain social circumstances provide the proper conditions for the correct idea of science to flourish.

All the same, not everyone will take this appeal to the criteria of science that just happen to be internationally current, seriously. In the first place, one could ask how generally accepted the empirical-analytical criteria actually are. The many methodological styles in the international social sciences that even today distance themselves from empirical-analytical thinking – modern hermeneutics and phenomenology, constructionism, ethnomethodology – are in this manner excluded without any further argument. Moreover, it is not at all certain that such a thing as "the" empirical-analytical methodology exists. In order to combat that idea, it is not necessary to be a Kuhn or a Collins. Laudan, who in many other respects explicitly distances himself from sociological science studies,[1] also repeatedly contends that methodological unanimity does not occur as a matter of course:

> One scientist, for instance, may believe (with Popper) that theory must make surprising, even startling predictions, which turn out to be correct, before it is reasonable to accept it. Another may be willing to accept a hypothesis so long as it explains a broad range of phenomena, even if it has not made startling predictions. A third may say (with Nagel) that no theory is worth its salt until it has been tested against a wide variety of different kinds of supporting instances. A fourth may demand. . . . All these familiar methodological principles of theory acceptance are at odds with one another, and each has found prominent advocates in recent science and philosophy.[2]

In this chapter, too, the position of international agreement on empirical-analytical rules fails to hold water. On closer philosophical consideration most Dutch psychologists will turn out to follow a quite individual variation within the international empirical-analytical framework. But not only that: This variation will appear to be directly related to the changing social identity of psychology in the Netherlands. The most important methodological rule of Dutch psychology, this chapter will conclude, is the product of the problems, unique to a certain extent, with which Dutch psychologists were confronted in the course of history and with which they would not have been confronted if their discipline had not had its own changing social role to play in its own changing society. In this way, finally, the standard perspective of a growing insight into an abstract general rationality will be exchanged for a different image.

Prediction in Dutch psychology

There is no science without predictions. It will have become clear in the preceding chapters that according to the current methodology of Dutch psychology this proposition closely reflects the importance of "prediction" in science. It was also repeatedly mentioned that above all it was the Amsterdam psychologist De Groot who in the 1950s began to espouse a concept of science in which prediction figures as a necessary requirement for a scientific psychology. Beginning in 1961 De Groot's ideas were further disseminated through his book *Methodologie*. Since then, not only have generations of Dutch psychologists but also generations of pedagogues, sociologists, politicologists, and even doctors and biologists have learned from one of the book's twelve Dutch editions how scientific research must be set up and carried out. For them "the methodology of De Groot" is a touchstone.[3]

Most practitioners of the human sciences in the Netherlands can therefore reproduce unerringly the various steps of "De Groot's empirical cycle": when you carry out research you begin with a stage of *observation* in which "empirical materials" are systematically collected and in which hypotheses are formed; in the second phase of *induction* you turn to a precise formulation of these hypotheses; next you proceed to the phase of *deduction,* which entails "the derivation of specific consequences from the hypotheses in the form of testable predictions," and you continue by *testing* these predictions against new empirical materials; finally, in the fifth and last stage, you proceed to the *evaluation* of the test results, after which, in the case of new related questions, the cycle begins over again from the beginning.[4]

This is the context in which it was instilled in Dutch researchers that prediction, as De Groot puts it, occupies *"a key position"* in scientific research. In supplementing the empirical cycle they also learned the much-quoted aphorism *"If I know something, I can predict something: If I cannot predict anything then I know nothing."*[5] Not only did Barendregt speak of "The Commandment: Thou shalt predict," but the methodologist/psychologist Willem K. B. Hofstee, at the beginning of his own book on social scientific methodology ("partly as a form of homage, and partly to prepare the rest of the discourse"), gives an extensive analysis of De Groot's aphorism.[6] Other Dutch psychologists quoted the saying, calling De Groot "the scientific conscience of Dutch psychology,"[7] and a Dutch introduction to psychology gives the aphorism, adding that it is "an almost classical" statement in science.[8]

Whether it is a question of people in general or diagnostic statements about individuals, De Groot always emphatically distinguishes not only theories and hypotheses, but also predictions. Although there are only differences of degree between hypotheses and theories, the line of demarcation between hypothesis and prediction is more clearly marked. De Groot's definition of prediction is: "A prediction refers to the expected outcome(s) of specific, i.e., pre-stated critical procedures, to be carried out on antecedently specified empirical materials . . . a prediction, in stating the outcome to be expected, refers to a specified sample or way of sampling."[9]

It is not necessarily a question of predicting *future* events. It is true that, according to De Groot, predicting the future, together with controlling and influencing it, is also an extremely important aim of science,[10] but that kind of predicting is not identical to the *methodological rule* of prediction. Predictions that serve to test hypotheses "may be made with respect to events in the past as well as in the present or in the future. . . . The justification of the term prediction (= foretelling) is that, in principle, the *verifying procedure* is situated in the future and has a still-unknown outcome, which can therefore be predicted. The outcome itself, however, may well be the result of events in the remote past."[11]

This methodological prediction requirement in the form of foretelling "future still-unknown *research outcomes*" is central to De Groot's book. He gives no methodological rules for the first phases of his empirical cycle: As far as the *invention* of hypotheses is concerned, there is "freedom of design." But these hypotheses must then be tested by predictions on new material. De Groot emphasizes this for hypotheses with regard to general laws and with regard to individuals, but equally to check the interpretation of one-off events and givens, such as clarifying the deeper meaning of a legend. That is possible, he argues, through the "partition" of the available data. In this, one part of the material, the legend, for example, is used to draw up hypotheses and predictions on the content of the other part. That other part then serves as the "new material" on which those predictions can be tested.[12]

Such predictions, it could be said, are mainly a means of allowing "the facts to speak" rather than the researcher. Predicting is supposed to protect researchers from a biased perception. Whoever draws up an explanation or interpretation runs the risk of projecting preconceptions. The derivation of predictions from explanations or interpretations of material not yet studied and the testing of these predictions is a means of controlling this. Anyone who, for example, claims that people having the astrological sign of Aries are pushy is not allowed to select a number of "Aries" and point out behaviors that reveal their pushiness. Even if this pushiness were so much greater in statistical significance than that of the control group of Aquarians, there is a big chance, as emphasized by De Groot's methodology, that random behavior was seen as a sign of pushiness. That is why the astrologist from the example should have predicted of a group of Ariens that in carefully specified situations they have behaved, or will behave, in a carefully specified manner, significantly differently. Or otherwise the researcher, on the basis of behavioral observations of a random group of people, should have been able to predict which of these were Ariens and which were not.

For De Groot, this prediction occupies a "key position," whereas in the "theory of social scientific research" of the psychologist and methodologist Hofstee, who was inspired by him, it is even proposed as *the king in the scientific game of chess.*"[13] Human behavior is much more changeable than the material on which the natural sciences make statements, Hofstee argues. Behavior can, for example, be changed by what is claimed about it or even only what is expected of it. As soon as social scientific statements are presented, their validity again becomes doubtful. An important consequence is that the validity of knowledge cannot be a criterion for distinguishing scientific researchers from non-scientific researchers. That criterion must therefore be the willingness of a researcher to run risks. And only by making concrete predictions about the expected results of concrete empirical research do you really run the risk of being put in the wrong.

However, that requirement in itself did not provide enough guarantees for Hofstee. After all, nothing compels a researcher to draw any consequences of displeasing research results. In this way he arrived at the proposal to radicalize the prediction requirement by means of the methodological obligation to make *bets.* The formulation of a prediction should imply making a bet (in the shape of a probability distribution) on the result of an empirical investigation. Anyone who comes up with ideas that are exciting enough will always find an opponent. "Scientific reputation," expressed in a points system, is what should be at stake.

Anyone making an empirical statement should thus accept the risk of loss of reputation. Nonprofitable theories, together with their proponents, would disappear naturally due to lack of reputation.

Every resulting betting *outcome* (and that is of crucial importance for Hofstee's theory), must then again be immediately conceived as a prediction (and therefore as a betting offer). For, as previously stated, Hofstee's basic principle is that it is impossible to say with any certainty of any tested social scientific statement that it will *remain* valid. The status of statements is and continues to be, therefore, that of a prediction. For the same reasons, for Hofstee it is more important to predict than to explain. An inevitable result of the principle that scientific knowledge can be made invalid at any moment is that it is not really important whether a prediction is related to an explanation: "If someone is crazy enough to come up with a prediction completely off the top of his head that is his affair."[14]

In this fashion De Groot's motto on prediction was translated, so to speak, by Hofstee into the proposition: *Because I never know anything for certain, I can only, and must constantly, predict.* From being a methodological key concept, prediction has become the be-all and end-all of methodology. If we add to this Hofstee's conviction that the social sciences are dependent on their methodology – they form "primarily a system to decide empirically differences of opinion on the question of how people act"[15] – then it follows that in his idea of science predicting concrete research results is in fact the hallmark of the social sciences as a whole.[16]

Prediction in logical empiricism

Logical empiricism arose in Europe in the 1920s and 30s, associated with authors like Carnap, Schlick, Reichenbach, and from a succeeding generation, Hempel, Nagel, and Feigl. It was this movement in the philosophy of science that was the chief source of inspiration for both De Groot and Hofstee. It is therefore obvious to make a comparison first with precisely this branch of international philosophy of science. In doing so not all similarities and differences will be traced, but only those that formed the core of the Dutch methodology, that is, the prediction requirement.

On examining the indexes of the relevant books, it is immediately striking that the term prediction is used far more by the Dutch.[17] A comparison on the basis of the meaning of the term shows that this quantitative distinction arises from qualitative differences.

In logical empiricism the concept of prediction is used in relation to what is called "the symmetry of explanation and prediction." Well-confirmed laws and theories that can be used to explain phenomena can also predict these phenomena. A proper explanation occurs when the appearance of the explained phenomenon could have been predicted on the basis of the theories and initial conditions used in the explanation. Whoever explains the sticking of a door by the fact that the door is made of wood and wood expands under certain circumstances can also predict when the door will jam. Similarly the concept of prediction is related to the *goal* of science. The logical empiricists also regard success on that level as a sign of the correctness of a theory. A special class of such predictions is formed by the predictions of "new facts": phenomena whose existence nobody suspected before or that did not play any role in drawing up the hypothesis. The ability to predict such new phenomena is the feature par excellence of a good theory.[18]

Even though the idea of the prediction of future events as the goal of science is also a matter of interest to Dutch psychologists and even though De Groot also points out in passing that "the man of science . . . is motivated particularly by a desire to be able to predict new phenomena,"[19] prediction in this sense does not express De Groot's basic idea of predicting future still-unknown *research outcomes*. That idea, after all, refers equally to predictions of events in the past. And when De Groot states that predictions must be tested on "new" *material*, by this he certainly does not mean that it must always be a matter of previously unknown *phenomena*. The requirement of prediction of research results refers also, and even primarily, to predictions which in that sense are not at all new, but that perhaps even played a clear role in drawing up the hypothesis. For example, the psychologist who wants to defend a theory on sex and intelligence, which implies men are more intelligent than women, must, according to the methodology of the Dutch psychologists, predict the result of intelligence measurements in samples of men and women specified beforehand. The predicted fact is then new in the sense that it is being tested or tested again on new material, but it is certainly not new in the sense that nobody had ever suspected its existence.

Naturally, the Dutch idea of prediction could also have been worded differently. In the famous book *Philosophy of Natural Science* by the logical empiricist Hempel we find the idea of the "test implication." Test implications, like the Dutch predictions, are derived from hypotheses. They state that under specified conditions C an event of kind E will occur.[20] At first glance similarity with the Dutch rule of prediction is large. However, the latter adds something essential to the notion of the test implication. The Dutch did not employ the concept of "prediction" by chance. The term makes it explicitly clear that the verifying procedure must always be situated in the future and must still have an unknown outcome when the test implication is formulated. A prediction refers, again to adduce De Groot's definition, "to the *expected* outcomes of specific, *prestated* critical procedures, to be carried out on antecedently specified empirical materials" (italics added). To put it another way, the concept of prediction further specifies *how to* check whether Hempel's event E occurs. The use of the term in this context indicates simultaneously how, methodologically speaking, a test implication must be handled.

Prediction in critical rationalism

There is an obvious explanation of this difference on the level of test instructions between logical empiricist philosophers and Dutch methodologists. What linked the logical empiricist philosophers was above all their attempt to give an epistemological justification of the empirical claims of the natural sciences; they analyzed the structure of existing scientific knowledge. For them the notion of explanation, and not that of prediction, was central. The Dutch psychologists, on the other hand, mainly designed rules for carrying out research. Whereas the logical empiricist philosophers were therefore chiefly interested in expressing the empirical foundation of theories, De Groot and Hofstee wanted to answer the question of how they themselves and their colleagues should work if they wanted their research to earn the qualification of being scientific.

The work of the Dutch methodologists shares this emphasis on questions of a methodological nature with that of the critical rationalist Karl Popper, who also ascribed an important role to prediction in his methodology. As a result, it is often stated that the methodology of Dutch psychology is a critical rationalist methodology.[21] Nevertheless there are crucial differences between the ideas of Popper and those of the Dutch psychologists, which will be examined now.

Popper's central idea is that it must be possible to subject theories to as sharp a critique as possible. His philosophy of science is therefore known as "falsificationism" because according to him scientific objectivity consists of the possibility of subjecting theories to falsification attempts. His falsificationist methodology means that it must be possible to derive from a theory risky predictions (and a number of specified initial conditions), which have the form of singular existential statements and which must be exposed to a confrontation with experience. If a positive decision is then made on the basis of empirical research about the acceptance of such a "basic statement," then the theory can be maintained. This raises the "degree of corroboration" of the theory.

In the fifth chapter of *The Logic of Scientific Discovery,* entitled "The Problem of the Empirical Basis," Popper expands on these basic statements and their testing.[22] In doing so he points out that researchers must arrive at "agreement" about research results. However, just as Hempel in his discussion of "test implications" did not indicate *exactly how* it must be checked whether something appears, Popper does not specify any additional rules for the way in which this "agreement on basic statements" must be brought about. The Dutch rule that the testing of predictions unconditionally consists of stating in advance still unknown research outcomes is not to be found in Popper.

However, as I have noted, Popper does use the term prediction in the context of testing theories. In order to point out more sharply the difference in Popper's use of the term with Dutch usage, a distinction could be made between two methodological preconditions: prediction I and prediction II. Prediction I stands for the rule that it must be possible to derive testable statements from a theory and prediction II stands for the rule that the testing of these statements involves taking a binding decision on the expected outcomes of prestated critical procedures to be carried out on antecedently specified empirical materials. Popper then writes only about prediction I; De Groot and Hofstee link both prediction I and prediction II to the concept of science.

This difference is not coincidental. That Popper, who also sets himself the goal of formulating methodological rules, does not specify in advance how basic statements must always be objectively tested and thus leaves open the possibility of agreement on a basic statement that has been created without being preceded by predictions of concrete research results[23] is not just carelessness on his part. It is a consequence of an important epistemological difference with the Dutch approach.

Popper developed his ideas in opposition to the logical empiricists. He considered their view, in which scientific knowledge is characterized by its basis on empirical evidence, to be incorrect. All knowledge, he argued, is always fallible knowledge. A clear boundary between scientific and nonscientific theories cannot be indicated. Popper's first argument for this was the "problem of induction": Logically speaking, the truth or increased probability of general claims can never be concluded from a finite number of statements. The appearance of a counterexample can never be ruled out.

His second argument, and that is what most concerns us here, was that the emergence of pure observational statements depends on a fiction.

The apparently purely empirical observational statements that formed the rock bottom of the logical empiricist view are also inevitably of a theoretical nature. A difference in principle between theoretical and empirical concepts cannot be defended, in Popper's view. He argues that every empirical statement uses concepts that go beyond the sensory experience of the moment. Empirical statements can therefore never be justified; objectivity and certainty about this are in principle not obtainable: "We do not attempt to justify basic statements. . . . Experiences can *motivate* a decision and hence an acceptance or a rejection of a statement, but a basic statement cannot be justified by them no more than by thumping the table."[24] That is the reason why Popper speaks of "the taking of decisions about basic statements" and not about the verification of observational statements. For Popper experience can serve only to motivate these decisions.

De Groot in particular, as far as this is concerned, is much more of a logical empiricist. He does subscribe to the problem of induction (Popper's first argument), but for him there is certainly a clear distinction between theoretical language and observational language. "Empirical concepts," he writes, "are derived from observational data by direct abstraction; hypothetical constructs, on the other hand, are inferred by reasoning or 'hypothesized.' "[25] Another characteristic of a prediction is that it "is formulated in such a way that it is strictly verifiable, that is, when put to the test it will prove either to be true or false."[26] Through the "verification of predictions," according to De Groot, hypotheses and theories are "anchored firmly to empirical fact."[27]

The case is somewhat different for Hofstee. He states emphatically that facts are conventions: "An operationalization is always arbitrary." However, at the same time he regards his betting model as a means to avoid "conventionalism": being put in the right is solely dependent on the degree of agreement between the real and predicted research results. Therefore, it is also true of this variation of what I call "Dutch prediction methodology" that an empirical statement based on "only consensus" plays no role in science. On the level of the facts, the Dutch teach, unlike Popper, that only the predicted results of research are decisive.[28]

Prediction in social-scientific methodology books

It would appear that the "key concept" or "the king in the scientific game of chess" in the methodology of Dutch psychologists is not to be found as such in the standard works on philosophy of science. There remains one other possibility, though. It could be that the Dutch prediction idea is a "natural implication" of the practical application of philosophical thought. Possibly this prediction rule appears to be inevitable as soon as the standard philosophy of science is further elaborated for everyday scientific practice. Even though the average textbook on social-scientific methodology, compared to the epistemologically based methodological reasoning of the Dutch, has in general more the character of a technical manual, perhaps this nevertheless is the place where the central idea of prediction must be looked for.

The study of a series of well-known books on methodology[29] reveals that the term "prediction" is used with diverse meanings. However, in none of the books studied is the concept of prediction related to the Dutch basic rule. Either they wonder whether or not prediction should be the goal of social science, or defend this position, or state that predictions must be based on an explanation, or else the question of the symmetry of explanation and prediction is dealt with. The concept of prediction is also used in the context of the validation of tests and other instruments, or it is noted that a statement's capacity to predict offers support to a theory and sometimes even that theories must be rejected by means of predictions. But in *none* of these works is the rule of investigating test implications or basic statements via prediction of research outcomes mentioned as a basic of science as it is in Dutch psychology.

Nor is the Dutch prediction rule made the core of the methodology under a different name. Each of the books studied emphasizes the importance of carrying out experiments, of objectivity, and of the operationalization of concepts, but each does so without making links to an unconditional requirement of specifying beforehand expected research results as a rigorous control of biased perception, not to mention building up the rest of their arguments around this. That also holds for the much-discussed technique of testing null hypotheses. Although it is true that in this instance a statement specified to a very concrete level is tested, this technique as such does not necessarily imply the rule that the data to be checked for significance can be gathered only after the careful formulation of a prediction.[30] In addition, the testing of null hypotheses is not in the same way the basis of a whole system of methodological rules, as the prediction of research outcomes is in Dutch psychology.[31]

In cases where the Dutch would emphasize the prediction rule, very different approaches may be presented for deciding the tenability of an empirical statement. A. Kaplan's famous book, *The Conduct of Inquiry: Methodology for Behavioral Sciences* (1964), in the section "The Process of Observation," gives the example of "observations of a child's behavior" that "except in very special circumstances, are inevitably colored by the emotional involvements with the child of those who have the most opportunity to observe him . . ." He does not then emphasize the Dutch prediction idea as a necessary remedy, but in one of his announced "general procedures by which errors of observation are taken into account" he mentions the idea that "the very *multiplicity of observers* may to some extent cancel out the effect of particular relationships" (italics added). And he adds another recommendation: "A very interesting compensatory device for the human factor is reported by Darwin, who tells us that he kept a separate notebook to record observations counter to his theory, lest he overlook or underestimate them."[32] The much-used *Scientific Method in Psychology* (1955) by C. W. Brown and E. E. Ghiselli says something comparable: "Eventually all predictions should be brought to *some kind* of empirical test. Empirical conditions . . . usually allow for the *coincident observation of the phenomena by many individuals,* and therefore increase the probability of reaching agreement among different investigators" (italics added).[33] To this the Dutch would comment that "some kind" of empirical test should always entail a prediction of research results, because the biased perceptions of "many individuals" might just add up to each other.

The occasional book in this area expresses the spirit if not the word of a prediction requirement that resembles the Dutch one. However, that happens only in passing and in a limited context, and is certainly not presented as the core of scientific thought in general. The German sociologist K. D. Opp, for example, states that a theory must be confronted with the facts by the quantitative formulation of potential falsificators, and on the basis of his example he too has a rule in mind that resembles the Dutch prediction requirement. However, Opp presents this only in the back of his book.[34]

We have now compared the prediction as it has come to be central in Dutch psychology via De Groot, with whatever rules are presented elsewhere in empirical-analytical-oriented manuals. The option that this prediction methodology was a natural and inevitable sequel of adapting abstract empirical-analytical philosophy of science for the practice of social scientific research turned out to be not tenable. The non-Dutch methodology books that were checked for this are not constructed in a comparable manner around the idea, so emphasized in the Netherlands, of predicting research results.

All of this philosophical hairsplitting served to carefully test the idea that only contemporary psychology in the Netherlands has the right to the epithet "scientific," *because* with its present-day empirical-analytical style it has finally conformed to the international standards of science. The possibilities of defending this idea now seem to be exhausted. First, if it is difficult enough to maintain that nowadays only empirical-analytical thinking is to be found in the international natural or social sciences and, second, if it is equally difficult to maintain that there is international consensus *within* empirical-analytical thinking, then present-day empirical-analytical methodology of Dutch psychology appears to mimic not even just *one* of the various international empirical-analytical schools. The difference may seem to be a difference in detail, but it is certainly not a minor detail: It concerns what the Dutch themselves regard as the "king" or "key concept" of social scientific methodology. Although it is true that this methodology belongs to the class of empirical-analytical thinking within which there are more mutual correspondences than there are with qualitative or interpretative-oriented thought, this methodology again confirms the already mentioned observation of other science researchers that the assumed unanimity on the rules of science is based on a fiction.

Furthermore, as shall now be explained, this specific core of Dutch psychological methodology is not a more or less *coincidental* variation. De Groot's prediction rule and its radicalization by Hofstee, as will be shown in the next two sections, are the products of unique developments that have taken place in the history of Dutch psychology. The circumstance that selection and vocational psychology was for a long time the biggest field of operation in psychology in the Netherlands – together with the social circumstances under which this field developed, which enabled *geisteswissenschaft* thought and phenomenology to gain a foothold – have contributed to the design of the present-day rules of the discipline. Partly by placing passages from the earlier chapters in a new context with regard to each other, and partly by supplementing them with new historical material, I will show how this prediction methodology can be made comprehensible on the basis of its *genealogy*. This characteristic feature of present-day Dutch psychology, I argue, can be traced back to the various branches of its family tree.

Intuition and prediction: The geisteswissenschaft *background to the Dutch concept of prediction*

In the works of Gerard Heymans, the concept of prediction appeared only in the context of predicting the future. And he thought it still far too soon for that: "The road which leads there, is that of the difficult and centuries-long ongoing collecting and processing of facts."[35] In his presidential address to the Eighth International Congress of Psychologists in 1926, this founder of Dutch psychology therefore warned: "We should abstain from prophesying, and learn to labor and to wait."[36]

The history of prediction as a methodological concept goes back to the 1920s when the first empirical-analytically trained Dutch psychologists began to operate in the field of psychotechnics. As shown in Chapter 2, these early psychotechnicians saw their task as discovering personal capacities in order to determine someone's suitability for a profession. The intention was to predict what someone would achieve in the future in a certain function. "In psychotechnics," the psychologist Prak commented, for example, "the intention is to predict the future by means of tests, as the testing of a metal for tensile strength also does."[37] This ability to predict the future as the goal of psychological activities would form the initial impetus for predicting research results.

The selection of people with the right capacities for certain professions was in the first place attempted by making an analysis of the profession, for which suitable staff had to be found. On the basis of such an analysis tests were designed that were expected to measure the necessary cognitive and sensorimotor capacities.

However, that in itself was not enough. Like the German–American psychologist Hugo Münsterberg, Prak and his colleague Brugmans also demanded that the tests be checked before being applied. Not only must the consistency of the test results when repeated with the same person be checked, but the connection between the test score and the expected work behavior also had to be investigated:

> The English and Americans have . . . terminology which we lack. They speak of reliability and validity. If the vocational advisor wants to conclude from the tidy appearance of the candidate that the young man is tidy in his work, then it must not be possible to disturb his impression by a casual encounter on the street when the Sunday suit is back in the wardrobe and the dirty jacket full of stains is back on, and it is necessary to first determine objectively that there is a correlation and how strong the correlation is between neat dress and accurate work.[38]

Mainly on the ground of the validity of their tests these early psychotechnicians emphasize the distinction between their selection practice and that of the layman. Professional psychotechnicians in their opinion must examine beforehand the extent to which their tests really show what they think they show.

In order to be able to accomplish this, a second assessment had to be available – repeatedly, in this context – alongside the test assessment, that could serve as a criterion to measure the value of the test. Brugmans and Prak stated that this had to be the judgment of the management of a company. They asked the employer for a "variation series" of the worst to the best worker. Only tests whose results showed sufficient agreement with the judgments given by the management could be used for selection purposes.

But the psychologists could not simply calculate the correlation between their own order and that of the management. That could too easily lead to willed results. "It has happened to me," related Brugmans, for example, "that a manager said to me: 'No sir, first *your* variation series; if I gave you my variation series first then I would make your research too easy." Since then I say that I must have the variation series, but that the people can be denoted as A, B, C, and so on. When the manager has then seen in which order, on the basis of psychotechnical research, Jansen, Pieterse, Bosma, Groenewoud, and so forth are placed, he can come with the file; Pieterse is A, Bosma is C, and so on."[39] If Brugmans thus on the insistence of a manager compelled himself to take an extra measure against possible subjective influences, he also occasionally declared "the desirability" of his own accord, "that only after the tests are finished the psychologist first observes which are the good workers and which the bad; because the more interest he takes in the tests, the more chance is there that he could – even if unconsciously – influence them in an undesirable and unacceptable manner."[40]

Thus there was a difference between Prak's measurement of the "tensile strength of a metal" and that of human capacities: The psychotechnicians in their empirical research were directly involved with co-assessing and skeptical managements, whom they had to convince of the value of their psychological judgment. In this context the idea emerged that it would be advisable to explicate one's own scores before looking at the criterion scores. In order to defend the value of the tests as convincingly as possible, the first test results had to be seen as an expression of an *expectation* with regard to criterion scores. In terms of contemporary methodology these test scores had to *predict* the criterion. It was in this context that "predicting" research results first emerged as a remedy against distorted perception.

Naturally this kind of predicting did not directly lead to the "birth" of the prediction methodology. The idea of the psychotechnicians concerned only the details of validating psychological tests and was certainly not presented as a primary methodological rule for social scientific research in general. Before reaching that stage, the history of Dutch psychology still had a long way to go. The next step was the implementation of *geisteswissenschaft* psychodiagnostics.

As also described in Chapter 2, the psychotechnicians attached to (mainly Christian) institutes in the 1930s had a more pedagogic view of their task. *Geisteswissenschaft* psychotechnicians assumed that every person is a unique individual and that the specific composition of some-one's capacities and motives cannot be studied by giving standard tests, but can be discovered only by approaching the individual as a unique person. Here candidates were asked to carry out tasks that did not serve to investigate to what extent they possessed an ability, but were, rather, a means of deriving information about their character from the *way* in which they approached their task. The so-called projection tests also helped the subjects to achieve "self-revelation." They were asked to tell what they saw in the presented figures or abstract ink spots. The answers served the psychologist as material on which to base interpretations.

Such tests were usually standardized to the extent that the accompa-nying manuals gave categories according to which the results had to be classified. However, their use was not preceded by research into the correlation between the test results and a criterion. That type of research was in conflict with the principles of the *geisteswissenschaft* psychodiag-nostics. After all when a test that is constructed according to correla-tional procedures is given, the subjects are compared with an earlier norm group of many from the "same species" and there is no space to reveal the unique aspects and abilities of a particular individual.

In the course of the 1930s and 40s the *geisteswissenschaft* approach increasingly gained ground. Soon after the Second World War just a few psychologists in the "elementarist tradition" were left. Of the first generation of empirical-analytical psychotechnicians only Prak continued to fight in all kinds of ways for such psychotechnics. In 1935 he founded his own Psychological Institute in The Hague and made study trips to the United States.

However, no matter how enthusiastic Prak was about the American methods, for research into someone's *charac.er* even he now called on "inborn capacity for empathy on the one hand, and extensive experience and practical knowledge of human beings on the other." In his eyes the correlational approach, concentrating on the measurement of sensorimo-tor and cognitive capacities, had little to offer on questions about charac-ter traits. With his opponents from the *geisteswissenschaft* school, Lun-ing Prak considered the American method of questionnaires, based on the calculation of correlation between answer patterns and character traits, to be unsuitable: "Such a list, particularly when contradictions like 'busily-lazy' and 'politely-unbearable' are placed on the list, often gives a better image of how the test subject would like to be or thinks he should appear, than how he really is! Naturally, with applicants you have to be particularly careful about this."[41]

Hesitantly, Prak therefore recommended interpretative techniques like the Rorschach test. He saw no other solution than to accept that "in the assessment of character it is necessary to be content with somewhat less objective certainty."[42]

It was this impasse that Prak's former employee De Groot attacked in the early 1950s in his new role as first assistant professor (*lector*) and shortly afterward as professor in applied psychology at the University of Amsterdam. De Groot made the methodological discovery of *combining* the idea – derived from the correlational approach – of a diagnosis as a prediction of a criterion, with the *geisteswissenschaft* doctrine that a person has to be studied in an interpretative manner "in his uniqueness."

This, too, was dealt with extensively in the second chapter: In his inaugural lecture of 1950 De Groot argued that "there can be no objection raised to a methodological development . . . of our intuitive understanding of our fellow man as such." However, the intuitively acquired statements should not be regarded as conclusions, but must be regarded as tools to arrive at *hypotheses from which testable predictions must be derived*. The way in which the interpretatively obtained insights must be studied for subjective elements was to make predictions about the behaviors of the subject.

In the succeeding years De Groot further developed his methodological ideas. He taught methodology at the Universities of Amsterdam and Nijmegen, and wherever possible put his proposals into practice. Not only Barendregt's research but also the experiences with the graphologists, which De Groot had first obtained as an employee in Prak's Psychological Institute in The Hague, strengthened in him the conviction that special methodological measures are needed to avoid biased perception.

In this period De Groot conceived the plan of writing a *book* on the basic principles of psychodiagnostics. Apart from English-language writers on test psychology such as Cronbach, Meehl, and Eysenck, he was inspired to develop his ideas further by various authors he had already become familiar with at an earlier stage. These included Von Mises, Reichenbach, Hempel, Feigl, and other logical empiricist philosophers of science, in which he also included Popper,[43] and the work of the German psychologist of thought, Otto Selz.[44] In this manner the manuscript gradually grew into a treatment of the methodology of much more than the original subject. So, in 1961 the first edition of *Methodologie* appeared: It had grown from a planned treatment of psychodiagnostics into a general manual on the "foundations of research and thought in the behavioral sciences."[45]

In this way prediction was elevated to a key methodological concept for psychology as a whole and for the social sciences in general. In *Methodologie,* translated for these sciences, we recognize the ideas that are derived originally from De Groot's combination of *geisteswissenschaft* and empirical-analytical psychodiagnostics. In *Methodologie* the ideas that intuition and fantasy can be applied, and that a person can be studied in his uniqueness, as long as this is constantly controlled by making concrete statements on behaviors to be expected have returned. They became the above-mentioned viewpoints that no methodological rules for formulating hypotheses are needed (one has "freedom of design" in this respect), and that science does not necessarily have to operate nomothetically (interpretations of individuals and legends also belong to scientifically researchable subjects), as long as all ideas are treated as predictions of research outcomes. "The criterion *par excellence* for 'true understanding' is in being able to predict the outcome of a test procedure."[46] Drawing conclusions is only justified after acquiring the outcomes of specific prestated critical procedures, to be carried out on previously specified empirical materials.

In the background to the creation of prediction methodology, therefore, psychodiagnostics for the benefit of selection and vocational psychology functioned as a prototype of empirical research. The specific meaning of and the strong emphasis on the prediction rule arose from the confrontation of the correlational and the *geisteswissenschaft* tradition, which was particularly fierce in Dutch psychotechnics. Anyone looking at De Groot's *Methodologie* from the angle of the debates from which it arose does not see a typical logical empiricist book with some critical-rationalistic elements, but an elaboration of individual viewpoints using whatever international philosophy of science has to offer.

Encounter and prediction: The phenomenological background to the Dutch concept of prediction

Not only early psychotechnics, but also postwar *phenomenological* psychology has left its traces in contemporary prediction methodology. It was the integration of the idea of self-responsibility from this psychology that ultimately led, in Hofstee's theory of science, to the elevation of prediction as the hallmark of a scientific attitude.

The earlier chapter on the Utrecht School described how personalistic-phenomenological psychology could arise precisely in the first postwar decades. Also discussed was the story of how the Utrecht professor Buytendijk, in his inaugural lecture of 1947, promoted the "encounter" as the way to understand the "innerness." Only through the "disinterested and desire-free yet personally interested participation in each other" can another person be really known. Whereas "natural-scientific psychology" turns people into objects and can lead only to superficial knowledge,

Figure 12. A. D. de Groot in front of the Amsterdam skyline, 10 November 1987 (photo copyright Klaas Koppe/Hollandse Hoogte, Amsterdam).

people are to be approached as fellow men in the understanding and open atmosphere of an encounter. Only in this way can psychology, the idea went, help people reach self-knowledge and liberate themselves from the inner obstructions that get in the way of a free and mature choice for the good.[47]

In the first postwar decades Dutch psychologists were even more divided among themselves than before. Whereas on the one hand many of them were now inspired by De Groot's methodology and English-language literature, others chose Buytendijk's encounter as a starting point. Van Lennep devoted the central part of his Utrecht inaugural

lecture to it. In this lecture Van Lennep contrasted the encounter with
both the empirical-analytical and *geisteswissenschaft* diagnostics. Much
less than he did in the 1930s, he would describe the *geisteswissenschaft*
research methods as superior to the "measuring" tests. Actually, a person is objectified in the *geisteswissenschaft* methods, too, Van Lennep
now stated.[48]

In Van Lennep's view objectification did not have to be scrapped
completely from psychology's repertoire, but after giving tests it was
necessary to have an encounter in which the objectification would be
counterbalanced. This would open the way not only to really discovering
someone's "innerness," but the opportunity would also be created to
make the test results obtained earlier available to the subject *himself*.
Only in that case could subjects themselves benefit from it: No longer
would they have to be determined by their character traits, but they
could transcend their characteristics. "I am not 'fixed,' " Van Lennep
said of such an imaginary subject, "in traits by the psychologist, but in
an image which offers me a guide in untangling the confused image that I
was to myself . . ."[49] In this way Van Lennep introduced a bipartite
division into the objectifying tests (both empirical-analytical and interpretative) and the nonobjective encounter.

In the work of the psychologist Benjamin Kouwer, who was also from
Utrecht, and who would later become Hofstee's most important teacher
in Groningen, these themes recurred in a developed and modified form.
In contrast to the "house philosophy" of the Utrecht School, which was
inspired by personalism, Kouwer was more in line with the *existential*
phenomenology of Sartre and de Beauvoir. As I have already explained,
Kouwer was convinced that people do not possess a "core," "soul," or
"essence" but are only that which they choose to do. In his view it was
therefore impossible to postulate absolute values.

When he became a professor in Groningen Kouwer stated in his inaugural lecture that on the basis of this idea, psychology had to do without
moral norms: "No matter how we search, a norm can never be given
which is a guideline for the advisory work of the practicing psychologist
as 'health' is for the doctor. Mental life cannot be led according to norms
fixed in advance. On the contrary, mental life is itself nothing else but a
constant *laying down* of norms."[50]

This nonexistence of given moral norms was reflected in two ways in
Kouwer's thinking about the task of psychology. In the first place he
emphasized that a psychologist had to protect himself with strict methodological measures from taking his own evaluations as facts: Kouwer
became a pioneer in the introduction and development of statistical techniques for validity research of psychological tests. However, in contrast

to many other psychologists who became engaged in test-research, he did not think that psychology could *restrict* itself to giving objective judgments. Like his earlier mentor Van Lennep, Kouwer believed that after research "objectification" had to be undone.

Nevertheless the second consequence of his existentialist emphasis on the absence of given norms was that the objectification should not therefore be continued in a veiled manner. And according to Kouwer that was the case in an encounter as defined by Buytendijk. Instead of the encounter, which still implied a form of studying, Kouwer introduced the notion of "the conversation." In a conversation an other is no longer studied, nor are norms and values imposed on the subject. The task of the psychologist after a diagnosis is merely to communicate the message that traits and aptitudes recorded with the help of tests could not be a norm for future behavior. The psychologists had to help people see the research outcomes as starting points for independently shaping their own existence.

The encounter whose purpose was to "really" get to know the "innerness" of the other and in which life-guiding help was offered, was thus replaced by Kouwer with the idea of an "open" conversation, in which Sartre's existentialist idea of "absolute freedom" was assumed and implicitly communicated. The objectification of the test situation was canceled in the sense that the earlier subject now, as recipient of the judgment, could transcend that judgment and if desired make it invalid; "With the help of the tests we can speak about what was hidden and dark, but what is spoken of here is past and dead."[51]

In general, as described in the earlier chapters, the norm-giving style of the older generation of psychologists had become increasingly less tenable and in the 1960s and 70s both the personalistic-phenomenological and the older *geisteswissenschaft* styles of thinking faded into the background. At the University of Groningen, Kouwer's staff and students developed, in his spirit, alternatives to the concrete norm-giving thought that was now seen as outdated.

One of these Groningen students was Hofstee, who after Kouwer's death in 1968 also became his successor. Hofstee approached the problems Kouwer had raised very much on the basis of the question of what this meant for the *scientific quality* of psychodiagnostics. Therefore, in Hofstee's book on psychodiagnostics in 1975, the two formerly parallel lines of the empirical-analytical and the existentialist-phenomenological thought came together more emphatically even than in Kouwer's writings.[52]

In that book Hofstee underwrote the importance of De Groot's prediction requirement. However, apart from that he stressed, following

Kouwer, that chiefly "behavior which in principle is changeable" must be the subject of psychological research and that predictions about such behavior must go to the persons involved themselves.

This combination of principles implies, Hofstee argued, that the predictions psychologists formulate as the basis of a psychological study can never have the status of conclusions. After all, as soon as a statement is made to persons about those persons, there is a large chance that the statement no longer has the same validity as it did formerly. Whether this concerns a general rule, or a judgment about an individual, a psychologist cannot help but to inform people what would have been the case if he or she had not informed them. Psychologists must make predictions during a study to check themselves, but after these statements have stood up to reality and therefore seem to be true, they soon revert to the uncertain status of a prediction because they are communicated to the subjects.

Six years after the publication of his book on pychodiagnostics, Hofstee's theory of social scientific research appeared. As I described, the problem of induction brought in from philosophy of science was the most important argument for seeing science at all times as a process of making predictions. However, whereas Hofstee argued that for the natural sciences this problem of induction "has about the significance of an abstract philosophical subtlety,"[53] for the social sciences it is a different matter. In the context of the social sciences *empirical* reasons can also be presented for distrusting generalizations about the future. These sciences, Hofstee argues, are concerned with changeable behavior and changeable cultural phenomena and furthermore the social sciences themselves contribute to those changes with their research results. For this reason social-scientific statements will never go beyond the status of a prediction, and it will therefore not be possible to defend a distinction between science and nonscience with an appeal to "knowledge." On this basis Hofstee developed the "betting model" already described.

So in Hofstee's work, the combination of De Groot's prediction idea and the existentialist-phenomenological theme of the encounter or the conversation in which people transcend what is said about them led to the radicalization described of the prediction concept in social-scientific methodology. Whereas De Groot's elevation of prediction to a key methodological concept was inspired by the fierce confrontation of correlational and *geisteswissenschaft* psychodiagnostics, in Hofstee the prediction, via a specific combination of existential-phenomenological and empirical-analytical elements into what was originally also psychodiagnostics, became the most important notion in the methodology of the social sciences.

In the methodology of Dutch psychology, and above all in its central idea that there is only science when future still-unknown research outcomes are specified, the very specific history of Dutch psychology appears to be given expression. In this the *geisteswissenschaft* and the phenomenological ways of thinking were not detours through which the development of the profession temporarily strayed from the actual "main road": They contributed to the formation of contemporary methodology.

Conclusions

If contemporary Dutch psychology has linked up with the criteria for science that happen to be the internationally valid ones, I argued at the beginning of this chapter, then the concept of "scientific" could be applied only to the current empirical-analytical psychology. Simply on empirical grounds, a clear boundary could be drawn between the current "scientific" and the earlier "nonscientific" psychology.

However, this was immediately refuted with the argument that, first, many alternatives to empirical-analytical thought are still defended in the international social sciences and, second, there are big differences even within the empirical-analytical style. The similarities within empirical-analytical thought seem to be based on family resemblances rather than on the presence of an important shared core. Thus the idea of internationally valid criteria for science is difficult to defend. In addition the "prediction methodology" of Dutch psychology turned out to offer a very specific variation and it also appeared that elements from other methodological styles have contributed to the design of the contemporary rules there.

On closer examination therefore it would appear not to be possible to exclude other constellations of norms by appealing to international agreement on "the" empirical-analytical criteria for science. It seems indefensible to deny the qualification "scientific" to phenomenological and *geisteswissenschaft* psychology, despite the fact that they were subscribed to by academically trained professionals and despite the fact that they – no less than the empirical-analytical methods – have turned out to be comprehensible and purposeful for particular goals and under particular circumstances.

This also implies that the *historical development* of methodological criteria cannot be simply interpreted as an ongoing process of scientific growth. The depiction of history in which the first empirical-analytical psychologists set the profession on the right track, in which *geisteswissenschaft* and phenomenological psychology derailed it again, and in

which finally the correct scientific way was followed only with the implementation of prediction methodology cannot be justified by appealing to
what is generally understood as science.

The preceding chapters, to recall them again, also showed that these
methodological changes did not appear *just like that*. They illustrated the
thesis that methodological changes are the result of the fact that the
profession, by keeping up with a changing society, changed itself. Methodology, as is constantly noted, expresses simultaneously both the scientific identity of psychology and its social identity. At times when the
social identity is no longer perceived to be tenable, "first" by the psychologist themselves or "first" by their public, the reigning methodology
comes under pressure and other criteria are sought.

The social identity of all methodological styles was not spelled out
again in the fifth chapter, but the idea that a methodological style comes
under pressure when the social identity connected to it is no longer
acceptable, also applies to those methodological changes that turned
out to be important for the emergence of prediction methodology. That
geisteswissenschaft thinking came to dominate the psychodiagnostics of
the 1930s is comprehensible only in terms of the pedagogic task carried
out in the institutions for psycho-social care, where psychodiagnosticians
seeking social progress generally worked. That empirical-analytical
methodology could prevail in psychodiagnostics in the 1950s and 60s was
connected to the decline of the harmony ideal seen between employers
and employees, which the *then* modernization-bound psychologists with
their methodological criteria sustained and contributed to. That the
existentialist-phenomenological philosophy of absolute freedom was in-
corporated methodologically shortly afterward was an expression of the
more general aim to exchange moral rules for individual responsibility.
In short, wherever there had been changes in the rules of the discipline
that led to the current prediction methodology, psychology and its methodology had kept apace with – or even actively contributed to – changing
social relations.

Therefore it can now be concluded that the alternative option on the
role of social changes, which was mentioned at the beginning of this
chapter, is not tenable. If the Dutch psychologists have, with their prediction methodology, linked up with internationally accepted criteria, the
idea went, then the changes in Dutch society have only worked to create
the preconditions for the establishment of the proper methodology. This
book would then just have shown that certain social conditions are necessary before a discipline can develop into a real science. But now that the
image of an existing universal methodology has turned out to be untenable, the other option on the role of social changes becomes more convinc-

ing. This is that a certain image of what may be called science can establish and maintain itself only when the ideas on social relationships it gives expression to are acceptable to its most important reference groups. To come back to the question of how it is that Dutch psychology experienced methodological changes of style in the course of time, the perspective of a growing insight into an abstract general rationality is ultimately exchanged for a different picture. That is the view of social scientists who, though making use of methodological ideas that are available elsewhere, nevertheless participate actively in their own society and in accordance with this define the rules – possibly local ones – of their discipline.

Epilogue: Social and rational rules

Other researchers have come to roughly similar conclusions on psychology's *practices* as I did on its *good intentions*. Danziger, for example, argues that the rise of the statistical approach in early American applied psychology reflected the ideologies in the American administrative contexts of education and the military. In a comparative section on the methods of psychology of expression (*Ausdruckspsychologie*) that flourished in German military psychology in the 1930s, he argues that these methods were geared to the Prussian ideas on "character" and "will" that were supposed to show themselves in facial expressions in stressful situations, such as a taut and upright position and a particular way of shouting commands.[1]

In contrast to the history Danziger has described, however, the present project has not focused on what researchers actually do but on what prominent representatives of the field write down that they *should* do. I have argued that not even these "statements of principle" are issued by scholars who, working independently of the time and culture in which they live, know what the rules of science are. If researchers in everyday practice always behaved according to their ideal image of science, the implication is, then that would still not mean that their approach could be culturally independent. That science and culture are not separated cannot be simply ascribed to *violation* of the rules.

Naturally, with this analysis of episodes from psychology in the Netherlands the role of social ideals in the establishment of a particular methodological style has by no means been exhaustively researched. On the other hand, there would appear to be no special reasons to assume that Dutch psychology is an *exception* with regard to the manner in which its methodological changes emerged. There seems to be no indication that the mechanisms described are characteristic of psychology in

153

the Netherlands only and that the plausibility of the image given could not be demonstrated with regard to methodological developments elsewhere in the social sciences. Looked at in this way the findings of this project do not merely affect the discussion of the status of Dutch psychology and many other human sciences in the Netherlands that are oriented to the methodology of the psychologists, but they affect the social sciences in general – which explicitly involve people and their mutual relations.

In the first chapter I explained my reasons for not taking a position in the debates that I intended to study. To put it very briefly, those reasons boiled down to the fact that there has already been much normative discussion about the correctness or incorrectness of methodological styles, but little research has been devoted to the empirical question of how these differences of style came into being. It is difficult to answer such an empirical question with an approach in which the researcher does not separate the participant's perspective from that of the reporter. This became very apparent in the historical chapters. For instance, as discussed in Chapter 3 initially I could not find any methodological program in the writings of the Utrecht phenomenological psychologists because I unwittingly was still taking a position *within* the methodological debates. In my search for the program of these phenomenologists, I was looking for alternative answers to the issues of empirical-analytical methodology, thereby inadvertently elevating the agenda of the empirical-analytical party to the standard. In this way, I would never have been able to discover the phenomenological program with its personalistic inspiration. Discovering the personalistic inspiration of the Utrecht phenomenology in its turn appeared to be an indispensable condition for finding an answer to my question of why this phenomenological psychology could flourish precisely in the early postwar years in the Netherlands.

The position of the methodological outsider, however, does not require *all-round and eternal aloofness*. The need for symmetry in the historical approach to the past does not have to lead to skeptical indifference. On the contrary, as George W. Stocking already put it at the end of his famous essay, "On the Limits of 'Presentism' and 'Historicism' in the Historiography of the Behavioral Sciences" (1965): "I would argue that the utilities we are seeking in the present are in fact best realized by an approach which is in practice . . . 'historicist.' " And: "By suspending judgment as to present utility, we make that judgment ultimately possible."[2]

However, whereas symmetry does not imply indifference, it also is not necessarily the historian him- or herself who must express the lessons for the scientific area studied. New insights yielded by a symmetrical approach, such as the idea that methodological styles differ according to the social identity they give expression to, can also be offered to the parties involved who can use them as they wish. After all, studying the historical development of a discipline does not yet make one an expert *in* that discipline. Serving up the consequences from an outsider's position can soon degenerate into pedantry. In this epilogue therefore I will confine myself to discussing briefly some respects in which the use of the nonnormative approach leads to normative matters.[3]

First, someone who does not participate in the methodological debate can still participate in a discussion *about* the debate. Where it concerns the metaquestion of how it happened that methodology changed, I already reached a conclusion that conflicts with that of the majority of the scientists studied. Insofar as psychologists have different views about how the variability of methodology must be understood, I was in fact regularly in debate with them. The main point that I contradicted is that only one of the styles discussed could justifiably claim the qualification "scientific."

The most important arguments in favor of this position that I have challenged are that former methodological styles were merely mistakes (Chapters 1–4) and that contemporary methodology in Dutch psychology simply reflects the universal standards of science (Chapter 5). I did this by replacing presentist accounts with contextualist accounts and by contesting the idea of worldwide unanimity on methodology.

Presentist rejection of the past and the supposed universality of contemporary methodology, however, are not the only arguments ever presented in defense of the idea that only one valid view of science exists. Attempts have also been made to base the superiority of a particular methodological style on the working methods used by successful scientists. The logical empiricist philosophy of science, for instance, is based on this. However, on closer examination this, too, failed to offer firmer ground. After careful empirical study successful researchers often appeared to be less doctrinaire than was originally thought. In addition, it could not really have been an argument, for methodological rules themselves define who can *count* as a successful scientist. Psychologists oriented to an empirical-analytical approach referred to other predecessors as successful models than their colleagues who were oriented to a phenomenological approach. Therefore, this strategy cannot point out either the single best methodology.[4]

Another current foundation of universal methodology turns out to be unstable. That is the foundation provided by the *goals of science* that are supposed to be served better in one way than any other. This cannot be

defended as a bastion of universal reason either. First, there is no basis
on which one of the many intended goals can be declared the only right
one, and second, decisions as to a particular goal can at best *eliminate*
some methodological procedures, but for the rest cannot unambiguously
dictate methodological decisions. Once agreement on goals is reached,
various plausible manners of attaining them often remain. That also
became apparent in the preceding chapters. Many paths were open both
with regard to what psychologists had to do, and *how* they had to do it.[5]

This would seem to mean that we have exhausted the possibilities for
rationally choosing one particular view of science, applicable in all cases.
The unavoidable conclusion would seem to be that the foundation of the
norms of methodology was formed by other norms and by nothing more
than norms. Along with methodological differences or along with the
rejection of a particular methodology as "unscientific," there were always – debatable – social assumptions.

Does this mean that every footing has disappeared, that in principle no
meaningful debate on what is to be called science is possible? Can one
choose only between methodological styles on grounds of taste? To my
mind, the finding that a methodological debate is at the same time a social
debate does not imply that only chaos would result or that discussions
can be won only by the most powerful party. That methodology has its
roots in the social ideals that practitioners of a discipline want to implement or support certainly does not imply that every methodological style
is equally good or bad, equally reliable or unreliable.

When in a concrete case agreement on methodology cannot be reached
via agreement on the direct aim of a project, then the larger framework
in which the project has to be carried through can become a subject of
debate, and, if necessary, this debate can be extended to the level of the
social identity of the discipline. That this is not an unreal picture was
also clearly demonstrated in the developments I have described. When
there was methodological disagreement, regularly there was also *open*
discussion on the social identity that psychology has or should have.
When a group tried to implement a new methodological style, the psychologists themselves often came up with overt sociopolitical arguments.
The *geisteswissenschaft* psychologists of the 1920s and 30s, the postwar
empirical-analytical psychologists who opposed the *geisteswissenschaft*
approach, the phenomenologists and those who displaced phenomenology with existentialist or empirical-analytical arguments with their battle
over methods were all at the same time involved in either proposals for
social change or pleas to maintain existing relations. If agreement on the
social identity of the field could not always be attained, at least clarity
did result in the divergent social positions, whereas if an appeal was

made to the supposed universal validity of criteria, then these positions remained implicit and unknown.

In this way, the loss of a solid basis for subscribing to one particular methodological style does not exclude the possibility of critique or debate. On the contrary. When the social side of methodology is acknowledged, it can become an *explicit* part of methodological debates. What Clifford Geertz has said about cultural differences in general also holds for scientific methodology. Without "the achievement of seeing ourselves amongst others, as a local example of the forms human life has locally taken, a case among cases, a world among worlds . . . objectivity is self-congratulation and tolerance a sham." Moreover, as history has amply showed, it is not even possible to exclude a methodological style by posing a universal definition of science. With equal justification – or rather equal lack of justification – the proponents of the approach under attack just put forward their own definitions. It seems to be precisely by an appeal to what happens to be real science, that discussions are simply short-circuited and that room for debate is wrongfully limited.

Appendix

List of people interviewed

H. Barendregt
J. Bastiaans
C. J. F. Böttcher
A. M. J. Chorus
L. Deen
F. J. P. van Dooren
J. J. Dijkhuis
J. H. Dijkhuis
N. H. Frijda
J. Groen
A. D. de Groot
H. J. A. Hamers
J. van Heerden

A. Jansen
I. A. M. H. van Krogten
M. J. Langeveld
J. E. van Lennep
M. van Loggem
H. van Luijk
H. Musaph
C. Sanders
J. T. Snÿders
R. Vuyk
W. Wijga
H. R. Wijngaarden

List of archives visited

Documentatie Centrum Nederlandse Politieke Partijen, Groningen
Internationaal Instituut voor Sociale Geschiedenis, Amsterdam
Katholiek Documentatiecentrum, Nijmegen
Rijksarchief Utrecht
Stichting Historische Materialen Psychologie, Groningen.

Notes

Chapter 1. The variability of methodological standards in the social sciences

1 Studies that support this notion are too numerous to mention. Some classics are: T. S. Kuhn, *The Structure of Scientific Revolutions* (Chicago: University of Chicago Press, 1962); P. Feyerabend, *Against Method* (London: New Left Bookstore, 1975); K. D. Knorr, *The Manufacture of Knowledge* (Oxford: Pergamon Press, 1981); G. N. Gilbert and M. Mulkay, *Opening Pandora's Box: A Sociological Analysis of Scientists' Discourse* (Cambridge: Cambridge University Press, 1984); Latour, B. and S. Woolgar, *Laboratory Life: The Social Construction of Scientific Facts* (Princeton: Princeton University Press, 1986). For a recent survey of the field see A. Pickering, ed., *Science as Practice and Culture* (Chicago: Chicago University Press, 1992).

2 J. A. Schuster and R. R. Yeo, eds., *The Politics and Rhetoric of Scientific Method: Historical Studies* (Dordrecht: Reidel, 1986); T. Pinch, "Normal Explanations of the Paranormal: The Demarcation Problem and Fraud in Parapsychology," *Social Studies of Science* 9 (1979):329–48.

3 For a survey up to 1986, see the introduction of J. A. Schuster and R. R. Yeo, eds., *The Politics and Rhetoric of Scientific Method*. See also the journal *Studies in History and Philosophy of Science*.

4 L. Laudan, *Science and Hypothesis: Historical Essays on Scientific Methodology* (Dordrecht: D. Reidel, 1981); L. Laudan, *Progress and its Problems* (Berkeley: University of California Press, 1977). For a discussion of Laudan with the modern sociology of science see J. R. Brown, ed., *Scientific Rationality: The Sociological Turn* (Dordrecht: D. Reidel, 1984). Laudan later also added the divergent goals of science as a determining factor in methodological choices (L. Laudan, *Science and Values: the Aims of Science and Their Role in Scientific Debate* [Berkeley: University of California Press, 1984]).

5 For instance, see S. Shapin and S. Schaffer, *Leviathan and the Air-Pump: Hobbes, Boyle, and the Experimental Life* (Princeton: Princeton University Press, 1985); K. Danziger, *Constructing the Subject: Historical Origins of Psychological Research* (Cambridge: Cambridge University Press, 1991).

160

6 Of course, this is not to say that no research into social scientific methodology has been done at all. Some titles include: G. Böhme, "Cognitive Norms, Knowledge-Interests, and the Constitution of the Scientific Object," in E. Mendelsohn, P. Weingart, and R. Whitley, eds., *The Social Production of Scientific Knowledge* (Dordrecht: D. Reidel, 1977), pp. 129–43; K. Danziger, *Constructing the Subject;* D. A. Mackenzie, *Statistics in Britain 1865–1930: The Social Construction of Scientific Knowledge* (Edinburgh: Edinburgh University Press, 1981); some of the chapters in L. Krüger, G. Gigerenzer, and M. S. Morgan, eds., *The Probabilistic Revolution: Vol. 2 Ideas in the Sciences* (Cambridge: MIT Press, 1987); G. A. Hornstein, "Quantifying Psychological Phenomena: Debates, Dilemmas, and Implications," in J. G. Morawski, ed., *The Rise of Experimentation in American Psychology* (Yale University Press, 1988), pp. 1–35; S. J. Gould, *The Mismeasure of Man* (New York and London: Norton, 1981); D. A. Stout, *Statistics in American Psychology: The Social Construction of Experimental and Correlational Psychology, 1900–1930* (Ph.D. dissertation, University of Edinburgh, 1987); I. Hacking, *The Taming of Chance* (Cambridge, Cambridge University Press, 1990).

7 Some historical publications, which deal with methodology in the social sciences that is not derived from the natural sciences, are: J. Platt, "Weber's Verstehen and the History of Qualitative Research," *The British Journal of Sociology* 36 (1985):448–66; A. Métraux, "Der Methodenstreit und die Amerikanisierung der Psychologie in der Bundesrepubliek 1950–1970," in M. G. Ash and U. Geuter, eds., *Geschichte der Deutschen Psychologie im 20. Jahrhundert* (Opladen: Westdeutscher Verlag, 1985), pp. 225–51; T. Rzepa and R. Stachowsky, "Roots of the Methodology of Polish Psychology," *Poznan Studies in the Philosophy of the Sciences and the Humanities* 29 (1992):231–49. In *Constructing the Subject* Danziger deals relatively briefly with German psychology based on a *geisteswissenschaft* orientation (pp. 129–34 and pp. 165–78).

8 These criteria are mentioned by T. S. Kuhn in the Postscript to the second edition of *The Structure of Scientific Revolutions* (1970), p. 184.

9 B. Latour, *Science in Action: How to Follow Scientists and Engineers through Society* (Cambridge: Harvard University Press, 1987), p. 30.

10 The figures come from Centraal Bureau voor de Statistiek, Statistisch Jaarboek 1991 (Den Haag: SDU, 1991). Forty percent of the Dutch population lives in the two western provinces of North and South Holland, together called the "Randstad," which forms one big urban agglomeration. There the population density is about 1,200 people per km^2.

11 This gives the Netherlands a greater "psychologist density" than Germany or England, for example, and probably the United States too. This information is taken from P. J. van Strien, "Psychology in The Netherlands," in A. R. Gilgen and C. K. Gilgen, eds., *International Handbook of Psychology* (New York: Greenwood Press, 1987), pp. 324–46. Van Strien took some figures from F. M. McPherson, "The Professional Psychologist in Europe," *American Psychologist* 41 (1986):302–5. See also M. R. Rosenzweig, "Trends in Development and Status of Psychology," *International Journal of Psychology* 17 (1982):17–140.

12 A. J. C. Rüter, quoted by J. P. Windmuller, *Labor Relations in the Nether-lands* (Ithaca, New York: Cornell University Press, 1969), p. 40. General information on the Dutch and their country is also to be found in: J. Gouds-blom, *Dutch Society* (New York: Random House, 1967); A. Lijphart, *The Politics of Accommodation: Pluralism and Democracy in the Netherlands* (Berkeley: University of California Press, 1968; revised repr. 1975); E. Zahn, *Das Unbekannte Holland: Regenten, Rebellen und Reformatoren* (Berlin: Siedler, 1984).

13 This holds for Dutch psychology in general (for an analysis of references to non-Dutch authors by Dutch psychologists in the course of time, see P. J. van Strien, "De Nederlandse psychologie in het internationale krachtenveld," *De Psycholoog* 23, (1988):575–86; P. J. van Strien, "Die Rezeption der Deutschen Psychologie in den Niederlanden," in A. Schorr and G. E. Wehner, eds., *Psychologiegeschichte Heute* (Goettingen: Hogrefe, 1990), pp. 261–74.

14 On the question of how it is that psychology in the United States developed in this way, see K. Danziger's detailed analysis, *Constructing the Subject,* a book that also clearly shows how big the variation is even *within* the set of quantifying methods. On the methodology of American psychology see also G. A. Hornstein, "Quantifying Psychological Phenomena."

15 S. Shapin, "Phrenological Knowledge and the Social Structure of the Early Nineteenth-Century Edinburgh," *Annals of Science* 32 (1975):219–43; S. Shapin, "Homo Phrenologicus: Anthropological Perspectives on a Historical Problem," in B. Barnes and S. Shapin, eds., *Natural Order: Historical Stud-ies of Scientific Culture* (London: Sage, 1979) pp. 41–71; S. Shapin, "The Politics of Observation: Cerebral Anatomy and Social Interests in the Edin-burgh Phrenology Disputes," in R. Wallis, ed., *On the Margins of Science: The Social Construction of Rejected Knowledge* (Keele: University of Keele, 1978), pp. 139–78.

16 H. M. Collins, "Son of Seven Sexes: The Social Destruction of a Physical Phenomenon," *Social Studies of Science* 11 (1981):33–62 (pp. 44–5). H. M. Collins (1985; repr. 1992), *Changing Order: Replication and Induction in Scientific Practice* (Chicago: The University of Chicago Press), pp. ix and 5–6.

17 For English-language articles on other aspects of psychology in the Nether-lands, see P. J. van Strien et al, "Psychology in the Netherlands"; P. J. van Strien, "The Omnipresence of the Social Sciences in the Netherlands and its Religious Roots," *Cheiron Newsletter* 9 (1991):8–16; P. J. van Strien, "Transforming Psychology in the Netherlands II: Audiences, Alliances, and the Dynamics of Change," *History of the Human Sciences* 4 (1991):351–69; E. Haas, "The Northsea Triangle: The Reception of German and Anglo-American Psychology in the Netherlands," *Cheiron Newsletter* 8 (1990):41–7; E. Haas, "The Development of School Psychology in the Netherlands," in H. Carpentero and E. Lafuente, eds., *New Studies in the History of Psychology and the Social Sciences: Revista de Historia de la Psicologia Monographs* 2 (1992), pp. 83–90.

18 A highly critical and extensive discussion of this project is in S. J. Gould, *The Mismeasure of Man* (chapter 6). For a critique of Gould's approach to the history of psychology, see M. Sokal, "Introduction: Psychological Testing and Historical Scholarship: Questions, Contrasts, and Context," in M. Sokal, ed., *Psychological Testing and American Society 1890–1930* (New Brunswick and London: Rutgers University Press, 1987), pp. 1–20.

19 M. Derksen and P. J. van Strien, "Typology and Individuality: Heymans' characterology and its Historical Background," in H. Carpentero and E. Lafuente, eds., *New Studies in the History of Psychology and the Social Sciences.*

20 Wilhelm Dilthey was the most outspoken nineteenth-century representative of the viewpoint that experimental psychology has little to offer when one really wants to understand (*Verstehen*) human behavior. Dilthey argued that humans should not be studied or observed as objects but as individuals who have intentions and feelings, and whose behaviors have meanings. Just as in everyday life it is possible to understand other human beings in an intuitive way (see also K. Danziger, *Constructing the Subject,* pp. 166–8). These ideas were elaborated by, among others, L. Klages, T. Litt, and S. Spranger, who were widely read by psychologists in the Netherlands. The Dutch translation of *Geisteswissenschaften* or *geisteswissenschaftlich* is *geesteswetenschappen* or *geesteswetenschappelijk*. However, in this book I will use the original German words.

21 L. van der Horst, "Over de Methodiek der Psychologie," *Nederlands Tijdschrift voor Psychologie* I (1933):40–53. N.B. This journal began in both 1933 and 1946 at volume 1. The years 1933 to 1945 are indicated in this book in Roman numerals, and the years following in Arabic numerals. Also, in 1946 the title of the journal was changed slightly to *Nederlands Tijdschrift voor de Psychologie.* The background to the journal's new start in 1946 is that the German occupying forces had replaced the sitting editorial board of the *NTP* in 1943 with the collaborating Dutch psychologist J. van Essen, which led to the publication of articles that had a clear National Socialist bent in the years 1943–5. See T. Dehue, "Niederländische Psychologie unter Deutscher Besetzung, 1940–1945," *Psychologische Rundschau* 39 (1988):39.

22 Spearman's appeal was translated into Dutch by the editors. The Dutch title was "Oproep Inzake Persoonlijkheidsonderzoek" (*Nederlands Tijdschrift voor Psychologie* I [1933]:289–95).

23 *NTP* editors, "Naschrift," *Nederlands Tijdschrift voor Psychologie* I (1933):295–7 (p. 296). In checking other European journals I found the Appeal in the French *L'Annee Psychologique* (32 [1931]:911–15) and in an abbreviated version in the Swiss *Archives de Psychologie* (XXIV [1934]:158–60). The French board included the notice without any comment, whereas the Swiss board introduced the piece with a statement of hope that Spearman's appeal would be heeded and bear fruit.

24 See, for example, A. D. de Groot, "Naar een Crisis in de Toegepaste Psychologie," *Nederlands Tijdschrift voor de Psychologie* 5 (1950):464–79.

25 A. D. de Groot, *Methodologie: Grondslagen van Onderzoek en Denken in de Gedragswetenschappen* (Den Haag: Mouton, 1961), p. 20. In the English translation, *Methodology,* this aphorism is given as: "If one knows something to be true, he is in a position to predict; where prediction is impossible, there is no knowledge" (A. D. de Groot, *Methodology* [Den Haag: Mouton, 1969], p. 20). Because the Dutch social scientists learned to say it in the first-person

singular, I use here the more literal translation than that given in the English version of the book.

26 For instance, a group of authors from Amsterdam University, which named itself "the Holzkampgroup" (after the Berlin Marxist psychologist Klaus Holzkamp), published "De Holzkampgroep," *Psychologie en Marxisme: Een Terreinverkenning* (Amsterdam: SUA, 1977).

27 W. K. B. Hofstee, "De Weddenschap als Methodologisch Model," *Nederlands Tijdschrift voor de Psychologie* 32 (1977):203–17; W. K. B. Hofstee, *De Empirische Discussie: Theorie van het Sociaal-wetenschappelijk Onderzoek* (Meppel: Boom, 1980). More on this in Chapter 5.

28 E. E. C. I. Roskam, *Mathematische Psychologie als Theorie en Methode* (inaugural speech, Nijmegen) (Nijmegen: Dekker & Van de Vegt, 1972). E. E. C. I. Roskam, "Verklaringsmodellen. De Modelbenadering in Onderzoek, Theorie- en Begripsvorming," in J. G. W. Raaijmakers et al., eds., *Metatheoretische Aspecten van de Psychonomie* (Deventer: Van Loghum Slaterus, 1983), pp. 42–68.

29 Even opponents to the standard image of science may hold that proposing alternatives implies refraining from the claim to practice science. D. N. Robinson, for instance, pleads for "the application of tried and true unscientific methods of analysis to those psychological problems that are nomothetically inexplicable" (D. N. Robinson, "Science, Psychology, and Explanation: Synonyms or Antonyms?" in S. Koch and D. E. Leary, eds., *A Century of Psychology as a Science* [New York: McGraw-Hill, 1985], p. 73).

30 The metaphor of the family resemblances comes from Wittgenstein (L. Wittgenstein, *Philosophische Untersuchungen* [Frankfurt am Main: Suhrkamp, 1971], p. 57).

31 To mention some titles: T. S. Kuhn, *The Structure of Scientific Revolutions;* T. S. Kuhn, *The Essential Tension* (Chicago: The University of Chicago Press, 1979); G. W. Stocking, "On the Limits of 'Presentism' and 'Historicism' in the Historiography of the Behavioral Sciences," *Journal of the History of the Behavioral Sciences,* 1, (1965):211–18; L. Graham, W. Lepenies, and P. Weingart, *Functions and Uses of Disciplinary Histories* (Dordrecht: Reidel, 1983); S. Shapin and S. Schaffer, *Leviathan and the Air-Pump;* H. Kragh, *An Introduction to the Historiography of Science* (Cambridge: Cambridge University Press, 1987), particularly chapter 9.

32 See, for example, the introduction in J. A. Schuster and R. R. Yeo, eds., *The Politics and Rhetoric;* M. G. Ash, "The Self-Presentation of a Discipline: History of Psychology in the United States between Pedagogy and Scholarship," in L. Graham, W. Lepenies, and P. Weingart, eds., *Functions and Uses of Disciplinary Histories,* pp. 143–89; M. G. Ash and W. G. Woodward, "Introduction," in *Psychology in Twentieth-Century Thought and Society* (Cambridge: Cambridge University Press, 1987).

33 Ash argued that in scholarly history of psychology it really *is* a dead horse (see M. G. Ash, "The Self-Presentation of a Discipline"). However, this is true only if one reserves the qualification "scholarly history" for that part of contemporary historiography of psychology that avoids presentisms.

34 H. Butterfield, *The Whig Interpretation of History* (London: Bell, 1931; repr. New York: AMS Press, 1978).

35 For an analysis of the changing significance of the concept of experiment in psychology and a comparison of the meaning of the concept in other sciences, see A. S. Winston, "The Construction of Experimental Method in Introductory Texts: An Interdisciplinary Analysis" (Proceedings of the XXIII meeting of Cheiron America, Slippery Rock University, 20–23 June 1991). For different meanings of "experiment" in biology, see P. Faasse, *Experiments in Growth* (Ph.D. dissertation, Department of Science Dynamics, University of Amsterdam, 1993.).

36 The critique formulated here of presentist approaches is also expressed in the demand for "symmetry" as made in "The Strong Programme of Sociology of Knowledge" (see for, example, D. Bloor, *Knowledge and Social Imagery* [London: Routledge & Kegan, 1976], p. 5). Here, too, it is proposed that ideas that deviate from the present must not be dismissed as false tracks, and why people thought as they thought, as well as why people now think as they think, must be investigated.

37 A comparable critique of such approaches was formulated by G. H. de Vries (G. H. de Vries, "Internalisering en Externalisering: Het Ontstaan van Rechtvaardigingscontexten," *Kennis & Methode* 3 [1979]:13–41. In G. H. de Vries, *De Ontwikkeling van Wetenschap: Een Inleiding in de Wetenschapsfilosofie* [Groningen: Wolters-Noordhoff, 1985]), pp. 142–4. For a critique of thinking in terms of internal and external factors, see also B. Latour, *Science in Action*.

38 For example, see H. Kragh, *An Introduction to the Historiography of Science*, p. 47. Because of this problem others propose to make a distinction between types of presentism, such as between presentism that is "enlightened" or not. See, for example, G. W. Stocking, "On the Limits of 'Presentism' and 'Historicism' in the Historiography of the Behavioral Sciences"; R. Smith, "Does the History of Psychology Have a Subject?" *History of the Human Sciences* 1 (1988):147–79.

39 R. de Wilde and G. de Vries presented a similar alternative manner of creating a line. They use Wittgenstein's metaphor of the thread that derives its strength not from one fiber running through the entire thread (which therefore has no core or essence) but from a large number of fibers overlapping each other (R. de Wilde and G. de Vries, "De Constructie van Historische Continuïteit: Disciplinegeschiedenis en het Verleden van de Sociologie," in J. L. Peschar and W. van Rossum, eds., *Wetenschap en Technologie: De Ontwikkeling van het Wetenschapsonderzoek in Nederland* [Deventer: Van Loghum Slaterus, 1987], pp. 56–71 [p. 68]).

40 J. Platt, "The Social Construction of 'Positivism' and its Significance in British Sociology," in P. Abrams et al., eds., *Practice and Progress* (London: Allen & Unwin, 1981), pp. 73–87 (p. 85).

41 K. Danziger, *Constructing the Subject* (p. 13). The discrepancy between rules and practice is also extensively shown in M. J. Mahoney, "Psychology of the Scientist: An Evaluative Review," *Social Studies of Science* 9 (1979):349–75.

42 I. Hacking, "The Accumulation of Styles of Scientific Reasoning," in Dieter Henrich, ed., *Kant oder Hegel? Ueber Formen der Begründung in der Philosophie* (Stuttgart: Klett Cotta, 1981), pp. 453–65.

43 For an analysis of the theoretical structure embedded in statistical inference,

see K. Danziger, "The Methodological Imperative in Psychology," *Philosophy of the Social Sciences* 15 (1985):1–13. See also G. Böhme, "Cognitive Norms, Knowledge Interests, and the Constitution of the Scientific Object" and G. Gigerenzer, "From Tools to Theories: A Heuristic of Discovery in Cognitive Psychology," *Psychological Review* 98 (1991):254–67.

44 There is some empirical evidence that at least parts of the social sciences are relatively strongly oriented to methodological prescriptions. A. Winston examined the degree to which introductory texts contain a general discussion of "research methods" of the kind that is now common in psychology texts. He compared introductory texts published on psychology, sociology, biology, and physics over three time periods: 1930–9, 1950–9, and 1970–9. The percentage of texts with discussions of research methods increased from 50% to 90% of texts in psychology, from 25% to 70% in sociology, from 23% to 45% in biology, and from 16% to 30% in physics. In neither biology nor physics *laboratory manuals* were these discussions to be found. Winston concludes that "In physics, which psychologists traditionally take to be the model science, discussions of research method and definitions of experiments are generally absent" (A. S. Winston, "The Construction of Experimental Method in Introductory Texts").

45 S. Woolgar, *Science: The Very Idea* (Sussex: Ellis Horwood, and London: Tavistock Publications, 1988), p. 107.

46 Recently philosophers and sociologists of science have argued that it is a mistake to suppose that scientific experiments always serve to test theories. See, for instance, I. Hacking, *Representing and Intervening* (Cambridge: Cambridge University Press, 1983); P. Galison, *How Experiments End* (Chicago: The University of Chicago Press, 1987); H. E. le Grand, ed., *Experimental Inquiries: Historical, Philosophical, and Social Studies of Experimentation in Science* (Dordrecht: Kluwer Academic Publishers, 1990).

47 Of course this is not to say that methodology does not have a legitimating function in the natural sciences. On this, see, for example, J. A. Schuster and R. R. Yeo, eds., *The Politics and Rhetoric of Scientific Method;* G. N. Gilbert and M. Mulkay, *Opening Pandora's Box.*

48 D. T. Campbell, "The Social Scientist as Methodological Servant of Experimenting Society," in S. S. Nagel, *Policy Studies and the Social Sciences* (Lexington: Lexington Books, 1969), pp. 27–32 (p. 27).

49 I prefer the concepts of "clientele" or "consumers" to the more common "audience." Audience suggests too much passivity. Audiences look and listen, but do not participate. As I will argue, there are other parties *in* the game. Clientele, though, should not be associated only with those who actually pay the professionals, but with all kinds of groups whose interests are involved and who possess enough social power to have a voice.

Chapter 2. Handwriting and character

1 The Dutch Psychological Association *(Nederlandse Vereniging voor Psychologie)* existed from 1927 to 1955 (see L. K. A. Eisenga, *Geschiedenis van de Nederlandse Psychologie* [Deventer: Van Loghum Slaterus, 1978], p. 67).

2 H. Y. Groenewegen, "De Grafologie op de Psychologendag," *Nederlands Tijdschrift voor de Psychologie* 3 (1948):350–4 (p. 351).

3 C. J. F. Böttcher, "Psychologie en Grafologie," *Nederlands Tijdschrift voor de Psychologie* 4 (1949):53–6 (p. 56).

4 E. A. Hof, "Enquête over de omvang van het gebruik van grafologie in de bedrijfspsychologie," in Nederlandse vereniging voor bedrijfspsychologie, *Discussieverslag van de Behandeling der Prae-adviezen over de Grafologie* (Amsterdam: Nederlandse Vereniging voor Bedrijfspsychologie, 1954) serial no. 19.

5 An extraordinary professor *(buitengewoon hoogleraar)* was one who was engaged in occupations outside the university as well as within it. Extraordinary professors held a part-time chair, usually one or two days a week. In the Netherlands the title was abolished in 1986. It is familiar not only in the Netherlands, but also in countries like Switzerland, Germany, and Belgium. See F. van Steijn, "Part-Time Professors in The Netherlands: Old Wine in New Bottles?" *European Journal of Education* 20 (1985):57–65.

6 This extraordinary chair in graphology (see note 5) was abolished in 1957.

7 Articles appeared in *De Telegraaf, Het Algemeen Handelsblad, De Nieuwe Rotterdamse Courant, De Haagse Post, Het Parool, Mens en Bedrijf, Het Financiële Dagblad, De Rotterdammer, Intermediair,* and *Vrij Nederland.*

8 According to an article in the French newspaper *Le Monde* (18 January 1989), in France 70% of companies and personnel bureaus use graphology and the Societé Française de Graphologie is flourishing. On the popularity of graphology in France see also J. Gooding, "By Hand, by Jove," *Across the Board* (December 1991):43–7, in which it is related that the French "minitel service" even presents "An Introduction to Graphology." With regard to Germany, the psychological dictionary of Dorsch (F. Dorsch, *Psychologisches Wörterbuch, 10. Neubearbeitete Auflage* [Bern: Hans Huber, 1982]) describes graphology as an essential tool for personnel psychology, psychiatry, the personality study of dead people in historiography, and even in choosing a partner. In 1978 Rainer Doubrawa noted in West Germany (including West Berlin) that around 4,000,000 decisions were taken annually in the field of staff selection on the basis of handwriting analysis and that 70% of the big companies were using graphological advice in selecting staff (see R. Doubrawa, *Handschrift und Persönlichkeit; Eine Kritische Studie zur Grundfragen der Graphologie mit einer Grafometrischen Untersuchung an Älteren Menschen* [Frankfurt: Peter Lang, 1978]).

9 NVP, "Sollicitatie Code," *Personeelbeleid* 16 (1980):1–4. In this journal of the Dutch Association for Personnel Policy disturbing publications are still appearing to this day on the use of graphology in other countries (for example: W. Pisa, "Irrationele methoden voor werving en selectie," *Personeelbeleid* 27 [1991]:138–43). Such articles also appeared in the journals of fellow associations in England and America, where graphology has never been well established. See, for example, A. Fowler, "An Even-Handed Approach to Graphology," *Personnel Management* (March 1991):20–5 in J. Gooding, "By Hand, by Jove," *Across the Board.*

10 In the first half of this century in the Netherlands, the designation *zielkunde* (science of the soul) was often used instead of the word psychology. This term stressed the contrast with "psychology without a soul" and its accompanying "materialistic view of man."

11 Heymans wrote on this in German, as he often did. G. Heymans, *Die Gesetze*

und Elemente des Wissenschaftlichen Denkens (Leipzig: Verlag von Johann Ambrosius Barth, 1890).

12 See P. J. van Strien and J. Verster, "The Response to Fechner in the Netherlands: Heymans' and Fechner's Monism," in J. Brozek and H. Gundlach, eds., *G. T. Fechner and Psychology* (Passau: Passavia Universitätsverlag, 1988), pp. 169–78. On Heymans in early twentieth-century Dutch intellectual and social context, see P. J. van Strien, "The Historical Practice of Theory Construction," in H. V. Rappard, P. J. van Strien, and L. Mos, eds., *Theory and History: Annals of Theoretical Psychology, vol. VIII* (New York: Springer, 1992), pp. 149–228.

13 G. Heymans, "Presidential Address, 1926," in *Gesammelte Kleinere Schrifte zur Philosophie und Psychologie II* (Den Haag: Martinus Nijhoff, 1927II), pp. 360–8 (p. 365).

14 Heymans himself wrote in Dutch and German, but not in English. English-language work on Heymans, apart from the aforementioned article by P. J. van Strien and J. Verster, includes: T. T. ten Have, "Essentials of Heymans' Philosophy," *Synthese* 5 (1947):526–41; and P. J. van Strien, "Transforming Psychology in the Netherlands II: Audiences, Alliances, and the Dynamics of Change," *History of the Human Sciences*, 4 (1991):351–71.

15 G. Heymans, *Inleiding tot de Speciale Psychologie*, vols. I and II (Haarlem: Erven F. Bohn, 1929). For this reason, H. Eysenck dedicated his famous *Readings in Extraversion-Introversion, vol. I, Theoretical and Methodological Issues* (London: Staples, 1970) to Heymans.

16 The Dutch designation *psychotechniek* stems from the German *Psychotechnik*. The term originally derives from William Stern, but became known mainly through the work of the German-American psychologist Münsterberg (H. Münsterberg, *Grundzüge der Psychotechnik* [Leipzig: Barth, 1914]). Münsterberg made a distinction between the use of psychology for practical purposes and for explanatory psychology. In the Netherlands *psychotechniek* was used until the 1950s and 60s; in English-speaking countries it was known as "occupational psychology."

17 H. J. F. W. Brugmans and J. L. Prak, "Een Psychologische Analyse van de Telefoniste," *Mededeelingen van de Dr. D. Bosstichting* (Groningen: J. B. Wolters, 1921), serial no. 3.

18 See N. Rose, *Governing the Soul* (London: Routledge, 1989; repr. 1991); K. Danziger, *Constructing the Subject;* J. Morawski, ed., *The Rise of Experimentation in American Psychology*.

19 A. Metraux, "Die Methodenstreit und die Amerikanisierung der Psychologie in der Bundesrepublik 1950–1970," in M. G. Ash and U. Geuter, eds., *Die Professionalisierung der deutschen Psychologie im 20. Jahrhundert* (Opladen: Westdeutscher Verlag, 1985), pp. 225–52; U. Geuter, *The Professionalization of Psychology in Nazi Germany* (Cambridge: Cambridge University Press, 1992).

20 F. Roels, "Rede," in *Verslag der Bijeenkomst ter Herdenking van het Tienjarig Bestaan van het Gemeentebureau voor Beroepskeuze te Amsterdam* (Amsterdam: Christelijke Psychologische Centrale voor School en Beroep, 1930), pp. 6–15 (p. 11).

21 K. Danziger points out that the idea of "character" as an object of systematic investigation originates with J. Bahnsen, *Beiträge zur Berücksichtigung pädagogischer Fragen* (Leipzig, Brockhaus, 1867), and that the measurement of character gained ascendency particularly in the context of the German military (K. Danziger, *Constructing the Subject*, p. 241, respectively p. 170).

22 D. J. van Lennep and T. Kuiper, "Methods of Selection of Personnel Suitable for High Administrative Positions," in *Sixth International Congress for Scientific Management, London, July 15th to July 20th 1935* (London: P. S. King and Son, 1935). For the rest, Van Lennep published very little in this period (see Chapter 3). The description of his working methods that follows is based on a book that appeared later from and about the Foundation (D. J. van Lennep, *Psychotechniek als Kompas voor het Beroep* [Utrecht: De Haan, 1949]).

23 D. J. van Lennep, *Psychotechniek als Kompas voor het Beroep*, p. 67

24 D. J. van Lennep, *Psychotechniek als Kompas voor het Beroep*, pp. 65–7.

25 Recently an American Ph.D. dissertation has been written on Van Lennep's Four Picture Test, which includes some biographical data on Van Lennep himself: M. A. Bryant, *D. J. van Lennep: The Four Picture Test and Female–Male Differences* (Ph.D. diss., University of Tennessee at Knoxville, 1990).

26 D. J. van Lennep, *Psychotechniek als Kompas voor het Beroep*, p. 129.

27 In 1622 a work appeared by C. Baldo, a professor at Bologna: *Trattato Come de Una Lettera Missiva si Cognascano la Natura Qualita del Scrittore;* in 1778, J. C. Lavater, a clergyman in Zürich, published his *Physiognomische Fragmente,* which devotes much space to graphology.

28 Tj. de Boer, "Grafologie," *De Beweging: Algemeen Tijdschrift voor Letteren, Kunst, Wetenschap en Staatkunde* 6 (1910):41 52; H. J. F. W. Brugmans, *Psychologische Begrippen en Methoden,* vol. I (Haarlem: Erven F. Bohn, 1922).

29 S. V. Margadant, *De Wetenschappelijke Grondslagen der Grafologie* (Leiden: Nederlandse Uitgevers Maatschappij, 1949). For a more extensive and more internationally oriented survey, see H. W. Müller und A. Enskat, "Grundzüge der Graphologie," in R. Kirchhoff, ed., *Handbuch der Psychologie, 5. Band, Ausdruckspsychologie* (Göttingen: Verlag Für Psychologie, 1972), pp. 533–86.

30 And in 1982 it even had its twenty-eighth reprinting.

31 D. J. van Lennep, *Psychotechniek als Kompas voor het Beroep*, p. 80.

32 Van Lennep in an unpublished catalogue of the Foundation (Archives of the History of Dutch Psychology, Groningen).

33 In Dutch the word *prak* means "cheap hashed meal."

34 J. Luning Prak, *De Moderne Onderneming en haar Personeel* (Amsterdam: Kosmos, 1947) p. 81–2.

35 I do not know whether this happened in other countries. As far as Germany is concerned, M. G. Ash specifically notes that the adherents of "scientific graphology" were competitors of the empirical-analytical test psychology *outside* the university, whereas *geisteswissenschaft* psychology was the competitor *within* the university (M. G. Ash, "Die experimentelle Psychologie und der deutschsprachigen Universitäten von der wilhelmischen Zeit bis zum

Nationalsozialismus," in M. G. Ash and U. Geuter, eds., *Die Professionalisierung der deutschen Psychologie im 20. Jahrhundert* (Opladen: Westdeutscher Verlag, 1985], pp. 45–89). On American psychology, J. M. Reisman says that the graphologist R. Saudek indeed enjoyed considerable favor among psychologists in the thirties and forties, but that "psychologists, especially those in the United States, were reluctant to investigate handwriting analysis seriously, probably because it was a technique . . . readily exploited by persons outside the profession" (J. M. Reisman, *A History of Clinical Psychology* [New York: Irvington Publishers, 1976], pp. 159 and 218).

36 For a history of the NIPP and NIP, see T. A. Veldkamp and P. van Drunen, *Psychologie als Professie: 50 jaar Nederlands Instituut van Psychologen* (Assen: Van Gorcum, 1988).

37 This emerges from correspondence between the Association and the Institute (Archives of the History of Dutch Psychology, Groningen).

38. A. D. de Groot, "Een Experimenteel-statistische Toetsing van Karakterologische (Grafologische) Rapporten," *Nederlands Tijdschrift voor de Psychologie* 2 (1947):380–473.

39 S. V. Margadant, "Het Toetsen van Grafologische Rapporten," *Nederlands Tijdschrift voor de Psychologie* 3 (1948):293–302 (p. 295).

40 A. D. de Groot, "Nogmaals: Toetsing van Grafologische Rapporten: Een Antwoord aan de Heer Margadant," *Nederlands Tijdschrift voor de Psychologie* 4 (1949):78–81 (p. 80).

41 A. D. de Groot, "Naar een Crisis in de Toegepaste Psychologie," *Nederlands Tijdschrift voor de Psychologie* 5 (1950):464–79.

42 De Groot, "Naar een Crisis in de Toegepaste Psychologie," p. 467.

43 A. D. de Groot, "Boekbespreking van Max Pulver 'Intelligenz im Schriftausdruck,' " *Nederlands Tijdschrift voor de Grafologie* 1 (1950):57–8.

44 A. D. de Groot, *Het Object van de Psychodiagnostiek* (Amsterdam: Noord-Hollandse Uitgevers Maatschappij, 1950).

45 A. D. de Groot, "Het Object van de Psychodiagnostiek," p. 9.

46 A. D. de Groot, "Scientific Personality Diagnosis," *Acta Psychologica* 10 (1954):220–41. Another early presentation in English is A. D. de Groot, "Some Preliminary Remarks to a Methodology of Psychological Interpretation," *Acta Psychologica* 6 (1950):206–30.

47 J. Th. Snijders, "Het Wetenschappelijk Karakter van de Psychodiagnostiek" (part I and II) *Nederlands Tijdschrift voor de Psychologie* 6 (1951):269–84 and 411–38 (p. 278).

48 Minutes of Graphology Working Group (unpublished, consigned to the Archives of the History of Dutch Psychology, Groningen).

49 Minutes of the Graphology Working Group.

50 The following descriptions of the experiments are derived from A. Jansen, *Toetsing van Grafologische Uitspraken* (Amsterdam: Van Rossen, 1963). A translated and adapted version is A. Jansen, *Validation of Graphological Judgments: An Experimental Study* (The Hague and Paris: Mouton, 1973).

51 A. Jansen, *Toetsing van Grafologische Uitspraken*, p. 127.

52 A. Jansen, *Validation of Graphological Judgments.*

53 J. J. Wittenberg, in *Vrij Nederland,* 17 October 1964.

54 A. Oldewelt Domesse en F. Domesse, *De Wonderlamp der Grafologie* (Amsterdam: Boucher, 1962).

55 H. Oldewelt, *De Nieuwe Rotterdamse Courant,* 4 January 1964.

56 Werkgroep Grafologie, *Een Wetenschappelijk Onderzoek van Grafologische Uitspraken* (Den Haag: Nederlandse Vereniging voor Bedrijfspsychologie, 1963) p. 13.

57 H. F. Kuiper-Talma Stheeman, unpublished report on the history of the Foundation for Psychotechnics Utrecht, 1966 (Archives of the History of Dutch Psychology, Groningen), p. 25.

58 H. R. Wijngaarden (interview, 1985). At the same time, P. J. D. Drenth, another psychologist from the Free University, was also engaging in full-page polemics in the newspaper *De Nieuwe Rotterdamse Courant* on "prediction versus intuition" with the graphologist E. Wolters (*NRC,* 19 January 1963; 9 March 1963; 23 March 1963; and 6 April 1963).

59 Apart from the work by De Groot already discussed, there was B. J. Kouwer, *Tests in de Psychologische Practijk* (Utrecht: Erven J. Bijleveld, 1952), a book on testing in psychological practice that was an early attempt in a Dutch context in treating technical terms such as reliability and validity. Also dating from the 1950s are S. D. Fokkema, *Psychologische Beschouwingen over het Leren Vliegen en over het Onderzoek naar de Geschiktheid als Vlieger* (Ph.D. diss., Free University at Amsterdam) (Groningen: Wolters Noordhoff, 1954); R. W. van der Giessen, *Enkele Aspecten van het Probleem der Predictie in de Psychologie* (Ph.D. diss., Free University at Amsterdam) (Amsterdam: Swets & Zeitlinger, 1957); and P. J. van Strien, "Het Onbehagen in de Psychologische Praktijk," *Nederlands Tijdschrift voor de Psychologie* 13 (1958):366–465.

60 A. D. de Groot, *Methodologie: Grondslagen van Onderzoek en Denken in de Gedragswetenschappen* (Den Haag: Mouton, 1961).

61 Empirical-analytical oriented works from the 1960s are, apart from De Groot's *Methodologie:* J. P. van de Geer, *De Mening van de Psycholoog* (Haarlem: De Toorts, 1961); P. J. D. Drenth, *De Psychologische Test* (Deventer: Van Loghum Slaterus, 1966); J. Linschoten, *De Idolen van de Psycholoog* (Utrecht, Bijleveld, 1964); J. Barendregt, *Research in Psychodiagnostics, Record of Investigations* (Den Haag: Mouton, 1961); and P. J. van Strien, *Kennis en Communicatie in de Psychologische Praktijk* (Utrecht: Bijleveld, 1966).

62 NIP, *Documentatie van Tests en Testresearch* (Amsterdam: NIP, 1969).

63 In 1933 L. C. T. Bigot and Ph. A. Kohnstamm wrote the work *Hoofdstukken uit de Psychologie,* which was subtitled "A Book for Headmaster Candidates" (Groningen: J. B. Wolters) and which was reprinted in 1946 by the same publisher as *Leerboek der Psychologie.* Under the last title the book went through many new editions intended to reach a broader public than headmaster candidates; it was written by H. C. J. Duijker with various other authors (Groningen: J. B. Wolters, 1958, 1960, 1962, 1964, 1968, 1969, 1970, 1976, and

1981). That Klages was able to raise graphology to a science is mentioned in the 1964 edition (H. C. J. Duijker, B. C. Palland, and R. J. Vuyk), p. 255.

64 H. C. J. Duijker, B. C. Palland, and R. J. Vuyk (1968).

65 W. K. B. Hofstee, *De Empirische Discussie: Theorie van het Sociaal-Wetenschappelijk Onderzoek* (Meppel: Boom, 1980), p. 37.

66 G. Visser, *Profiel van de Psychologie* (Muiderberg: Coutinho, 1985), p. 64.

67 These figures are derived from the unpublished report, *Nederlandse Stichting voor Psychotechniek, De plaats van de Nederlandse Stichting voor Psychotechniek in Nederland*, 1946 (Archives of the History of Dutch Psychology, Groningen).

68 See, for instance, J. P. Windmuller, *Labor Relations in the Netherlands* (Ithaca, New York: Cornell University Press, 1969). According to some, the limited rebelliousness of the Dutch people dates back many centuries earlier. See C. R. Boxer, *The Dutch Seaborne Empire: 1600–1800* (London: Hutchinson, 1965, 1977).

69 See J. P. Windmuller, *Labor Relations in the Netherlands;* A. Lijphart, *The Politics of Accommodation: Pluralism and Democracy in the Netherlands* (Berkeley: University of California Press, 1968); and E. Zahn, *Das Unbekannte Holland* (Berlin: W. L. Siedler Verlag, 1984).

70 The term "pacification democracy" stems from A. Lijphart, *The Politics of Accommodation*. The development of the *sciences* in the Netherlands is also related to pacification democracy and the pillarization of Dutch society. German and English-language publications on this topic include (chapters and passages in): E. Zahn, *Das Unbekannte Holland;* P. J. van Strien, "Transforming Psychology in the Netherlands II: Audiences, alliances, and the dynamics of change," *History of the Human Sciences* 4 (1991):351–69; S. Blume, R. P Hagendijk, and A. A. M. Prins, "Political Culture and the Policy Orientation in Dutch Social Science," in P. Wagner, C. H. Weiss, B. Wittrock, and H. Wollman, *Social Sciences and Modern States: National Experiences and Theoretical Crossroads* (Cambridge: Cambridge University Press, 1991), pp. 168–91.

71 H. Daalder, "Parties and Politics in The Netherlands," *Political Studies* 3 (1955):1–16; H. Daalder, "The Netherlands: Opposition in a Segmented Society," in R. A. Dahl, *Political Oppositions in Western Democracies* (New Haven: Yale University Press, 1966), pp. 188–236; H. Daalder, "The Consociational Democratic Theme," *World Politics*, 26 (1974):604–21; J. Obler, et al., eds., *Decision-Making in Smaller Democracies: The Consociational Burden* (London: Sage, 1977).

72 For comparison with the United States, Great Britain, and Germany; see K. Danziger, *Constructing the Subject;* with Great Britain: L. S. Hearnshaw, *A Short History of British Psychology* (London: Methuen, 1964); N. Rose, *Governing the Soul: The Shaping of the Private Self;* with the United States: D. S. Napoli, *Architects of Adjustment: The History of Psychological Profession in the United States* (Port Washington and London: Kennicat Press, 1981); with Germany: U. Geuter, *Die Professionalisierung der deutschen Psychologie;* S. Jaeger, "Zur Herausbildung von Praxisfeldern der Psychologie bis 1933," in M. G. Ash and U. Geuter, *Geschichte der deutschen Psy-*

chologie im 20 Jahrhundert: Ein Ueberblick, pp. 83–113. France seems to offer another special case. Here too individuality was stressed and American methods were rejected. Psychotechnics was not so much used in the army, but in criminal justice, education, and vocational training (W. M. Schneider, "After Binet: French Intelligence Testing 1900–1950," *Journal of the History of the Behavioral Sciences* 28 [1992]:111–32).

73 H. J. F. W. Brugmans, "De Organisatie van de Arbeid: het Taylorsysteem en de Psychotechniek," *Mededeelingen van de Dr. D. Bosstichting* (Groningen: J. B. Wolters, 1923), p. 25 (serial no. 6).

74 F. Roels, *Psychotechniek van Handel en Bedrijf* (Amsterdam: De R. K. Boekcentrale, 1920), p. 136.

75 E. J. van Det, "Rede," (Amsterdam: Christelijke Psychologische Centrale voor School en Beroep, 1930) pp. 15–28 (p. 17). An exception was Taco Kuiper, co-founder of the *Nederlandse Stichting voor Psychotechniek.* Kuiper was a Communist and he saw psychotechnics as a way to abolish capitalism. However, he was not very satisfied with psychotechnics as it functioned in his day: "Whoever serves a bad master, becomes his slave." Kuiper's views on this can be found in T. Kuiper and D. J. van Lennep, "Psychotechniek," in H. J. Pos et al., eds., *Eerste Nederlandse Systematisch Ingerichte Encyclopedie* (Amsterdam: ENSIE, 1947), pp. 506–11 (serial no. I). According to a personal communication from Van Lennep's successor in the Foundation, W. Wijga, Van Lennep wrote the methodological section and Kuiper the social-critique parts.

76 H. T. de Graaf, "De Invoering van de Psychotechniek in het Sociale Leven," *De Opbouw: Democratisch Tijdschrift* 4 (1921–2):814–20 (p. 820).

77 G. J. van der Ploeg, *Vierde Jaarverslag van de Christelijke Psychologische Centrale voor School en Beroep* (Amsterdam: Christelijke Psychologische Centrale voor School en Beroep, 1929), p. 12.

78 F. Roels, "Rede," p. 13–14.

79 E. J. van Det, "Rede," p. 21.

80 T. Kuiper and D. J. van Lennep, "Psychotechniek," p. 508.

81 In Germany at this time there also arose a demand among employers for personal qualities rather than technical and cognitive skills and here too the American tests were rejected. Danziger (*Constructing the Subject,* pp. 169–70) explains this by pointing out that in Germany psychodiagnostics flourished mainly by offering its services to the army. Psychology had to speak the right language to grab the attention of the superiors and *Karakter* was a highly valued characteristic in the Prussian-German military tradition. Note that this explanation agrees with the one I have presented, in the sense that here the clientele of the profession and presumptions current among them are also referred to.

82 E. Zahn, *Das Unbekannte Holland.*

83 A. Lijphart, *The Politics of Accommodation: Pluralism and Democracy in the Netherlands* (Berkeley: University of California Press, 1975), revised repr. 1968 (p. vi).

84 J. P. Windmuller, *Labor Relations,* chapter 10. On social changes in the

sixties and seventies, see the chapter added to the Dutch translation by C. de Galan in J. P. Windmuller and C. de Galan, *Arbeidsverhoudingen in Nederland* (Utrecht/Antwerpen: Ambo, 1977), serial no. 2; and C. P. Middendorp, *Ontzuiling, Politisering en Restauratie in Nederland* (Meppel: Boom, 1979).

85 *Overleg* means something like organized consultation. However, it has often been observed that the term is specifically Dutch (E. Zahn, *Das Unbekannte Holland;* J. P. Windmuller, *Labor Relations in the Netherlands*). Windmuller wrote in 1969: "It is revealing that the term *overleg* not only means consultation but also refers to the collective bargaining process. The parties practice *overleg* for the purpose of writing a contract. *Overleg* has the connotation of peaceful, amicable, and unhurried discussion. There is a separate term for negotiating *(onderhandelen),* but there is no direct counterpart for the English and American expression of bargaining or collective bargaining with its implied meaning of haggling, horsetrading, and continuous contract administration" (pp. 436–7).

86 *Elseviers Weekblad,* 12 August 1950.

87 *Het Vrije Volk,* 26 February 1955; 5 March 1955; 12 March 1955; 19 March 1955.

88 *De Telegraaf,* 14 May 1966.

89 NVV, *Welvaartsplan van het Nederlands Verbond voor Vakverenigingen* (Amsterdam: NVV, 1952).

90 W. Top, "Naar de bliksem met de psychotechniek," *De Vakbeweging,* 64 (1970):6.

91 *De Telegraaf,* 14 May 1966.

92 See, for instance, C. Alexander, *Personality Tests: How to Beat Them and Make Top Scores?* (New York, 1965); L. Baritz, *The Servants of Power* (Middletown, CT; Wesleyan University Press, 1960; repr. Westport, CT: Greenwood Press, 1974); B. Hoffman, *The Tyranny of Testing* (London: Collier-McMillan, 1962).

93 P. J. van Strien, interview in the *Haagse Post,* 23 December 1969.

94 W. K. B. Hofstee in *Vrij Nederland,* 16 March 1974.

95 P. J. van Strien, ed., *Personeelsselectie in Discussie* (Meppel: Boom, 1976).

96 Sociale Zaken, *Een Sollicitant is ook een Mens: Eindrapport van de Commissie Selectieprocedures* (Den Haag: Staatsuitgeverij, 1977).

97 Just as there was a swing toward the interpretative approach in the methodology of German psychology between the two world wars, in the fifties there also was an empirical-analytical turnaround. Furthermore, as Métraux describes the latter was arising, as in the Netherlands, from validity problems in psychodiagnostics (A. Métraux, "Der Methodenstreit und die Amerikanisierung der Psychologie in der Bundesrepublik 1950–1970"). Métraux's article does not aim to explain this change, but rather to describe its institutional and cognitive aspects. Insofar as he searches for an explanation he indicates that a younger generation of psychologists wanted to dissociate itself from the former generation that had quite often worked for the Nazis. Besides, Métraux points to a growing need for the legitimation of psychology with regard to other sciences. Therefore, in this case it is still unclear whether the *clientèle* of the discipline also had a voice, and if so, how.

Chapter 3. *"Like everything living which encounters us"*

1 H. Misiak and V. S. Sexton, *Phenomenological, Existential, and Humanistic Psychologies: A Historical Survey* (New York and London: Grune & Stratton, 1973), p. 36.

2 H. Spiegelberg, *The Phenomenological Movement: A Historical Introduction* (The Hague M. Nijhoff, 1956; revised repr. 1982); H. Spiegelberg, *Phenomenology in Psychology and Psychiatry: A Historical Introduction* (Evanston, IL: North Western University Press, 1972); J. J. Kockelmans, *Phenomenological Psychology: The Dutch School* (Dordrecht, Boston, Lancaster: M. Nijhoff, 1987); G. Thinès, *Phenomenology and the Science of Behaviour: An Historical and Epistemological Approach* (London: Allen and Unwin, 1977).

3 Most of the works of this phenomenological psychology are written in Dutch. Many of them are translated into German or French and some into English. Some English translations are: F. J. J. Buytendijk, *Pain: Its Modes and Functions* (London: Hutchinson, 1961); F. J. J. Buytendijk, *Woman: A Contemporary View* (Glenn Rock, NY: Newmann Press and Association Press, 1968); F. J. J. Buytendijk, *Prolegomena to an Anthropological Physiology* (Pittsburgh: Duquesne University Press, 1974); D. J. van Lennep, "The Four Picture Test," in H. H. Anderson and G. L. Anderson, eds., *Introduction to Projective Techniques* (New York: Prentice Hall), pp. 149–80; D. J. van Lennep, "Projection and Personality," in H. P. David and H. V. Bracken, eds., *Perspectives in Personality Theory* (New York: Basic Books, 1957), pp. 259–77; D. J. van Lennep, *Guidelines for the Use of the Four Picture Test* (Lisse: Zwets & Zeitlinger, 3rd rev. ed. 1983); J. Linschoten, *On the Way Toward a Phenomenological Psychology* (Pittsburg: Duquesne University Press, 1968); translations of articles of the Utrecht group are to be found also in J. J. Kockelmans, ed., *Phenomenological Psychology: The Dutch School.*

4 This sense of amazement was noted by, among others, W. J. M. Dekkers, *Het Bezielde Lichaam: Het Ontwerp van een Antropologische Fysiologie en Geneeskunde volgens F. J. J. Buytendijk* (Zeist: Kerckebosch, 1985), p. 323; R. H. J. ter Meulen, "De Deugd van een Geboren Heer: F. J. J. Buytendijk over Deemoed, Ontmoeting en Seigneurale Cultuur," *Psychologie en Maatschappij* 11 (1987):5–23, p. 16; R. H. J. ter Meulen, *Ziel en Zaligheid: De Receptie van de Psychologie en de Psychoanalyse onder de Katholieken in Nederland* (Baarn: Ambo, 1988), p. 209; F. Schenk, *De Utrechtse School: De Geschiedenis van de Utrechtse Psychologie tussen 1945 en 1960* (Utrecht: Instituut voor Geschiedenis, 1982), p. 10.

5 Buytendijk's substantial bibliography of 391 titles, including the many translations of his works, is published in W. J. M. Dekkers, *Het Bezielde Lichaam* (p. 290–318). *The Mind of the Dog* was published in English, too (London: Allen & Unwin, 1935; repr. 1973).

6 That semiapologetic aspect was noted by Rutten, who also gave the quote from Buytendijk (F. Th. Rutten, "In Memoriam Prof. Dr. F. J. J. Buytendijk," *Nederlands Tijdschrift voor de Psychologie* 30 (1975):iii–xviii (pp. vi–vii).

7 As related by Rutten himself in F. Th. Rutten, "In Memoriam Prof. Dr. F. J. J. Buytendijk."

8 T. A. Veldkamp and P. van Drunen, *Psychologie als Professie: 50 Jaar Nederlands Instituut van Psychologen* (Assen: Van Gorcum, 1988), p. 12.

9 Source: Letters from the Faculty to the Board of Governors, State Archives Utrecht, archives of the "College van Curatoren" of Utrecht University, nr. 711.

10 Langeveld (1980), interview with F. Schenk.

11 D. J. van Lennep, *Psychologie van de Projectieverschijnselen* (Utrecht: Nederlandse Stichting voor Psychotechniek, 1948).

12 J. H. van den Berg and J. Linschoten, eds., *Persoon en Wereld: Bijdragen tot een Fenomenologische Psychologie* (Utrecht: Bijleveld, 1953).

13 M. J. Langeveld ed., *Rencontre, Encounter, Begegnung* (Utrecht: Het Spectrum, 1957).

14 F. J. C. Hogewind and C. Sanders, *Hoe Ontmoet Ik mijn Medemens?* (Utrecht: Bijleveld, 1958).

15 In 1964 Linschoten distanced himself controversially from phenomenology with the (posthumous) book *Idolen van de Psycholoog* (Utrecht: Bijleveld, 1964). I will devote more to this later.

16 Some recent titles advocating phenomenology in psychology are: A. Giorgi, *Psychology as a Human Science: A Phenomenologically Based Approach* (New York: Evanston and London: Harper and Row, 1970); D. Kruger, *An Introduction to Phenomenological Psychology* (Pittsburgh, Duquesne University Press, 1981); A. Giorgi, ed., *Phenomenology and Psychological Research* (Pittsburgh, Duquesne University Press, 1985); A. Ashworth, A. Giorgi, and A. J. J. Koning, eds., *Qualitative Research in Psychology* (Pittsburgh, Duquesne University Press, 1986); M. Hammond, J. Howarth, and R. Keat, *Understanding Phenomenology* (Oxford: Basil Blackwell, 1991).

17 The aforementioned historical surveys by Spiegelberg, Misiak and Sexton, and Thinès, for example, are clearly "insider" histories of phenomenology.

18 F. J. J. Buytendijk, *Het Kennen van de Innerlijkheid* (inaugural lecture at Utrecht) (Nijmegen and Utrecht: Dekker & Van de Vegt, 1947), p. 18.

19 Examples are to be found in the Catholic Documentation Center at Nijmegen, Buytendijk Archives, nr. 397.

20 Quoted from F. Schenk, *De Utrechtse School* (p. 28).

21 M. J. Langeveld, *Capita uit de Algemene Methodologie der Opvoedingswetenschap* (Groningen: Wolters Noordhoff, 1972), p. 73. In a discussion of De Groot's lecture of 1950 Langeveld already spoke of a "fatal conception." See M. J. Langeveld, "Bespreking van 'Het Object van de Psychodiagnostiek,' Oratie door A. D. de Groot," *Paedagogische Studieën* 27 (1950):381–2. De Groot's ideas were also emphatically rejected in M. J. Langeveld, *Over het Wezen der Paedagogische Psychologie en de Verhouding der Psychologie tot de Paedagogiek* (Groningen: J. B. Wolters, 1951).

22 F. Th. Rutten, "In Memoriam Prof. Dr. F. J. J. Buytendijk," p. xiii.

23 F. Schenk, *De Utrechtse School*, pp. 13 and 17.

24 The quote comes from J. Beijk, "De Cognitieve Revolutie in de Psychologie: Deel 1," *De Gids* 143 (1980):317–32 (p. 328).

25 J. Linschoten, "Nawoord," in Van den Berg and Linschoten, *Persoon en Wereld*, pp. 252–53.

26 Kuhn gave a graphic description of such a gestalt-switch in his attempts to understand Aristotelian mechanics (T. S. Kuhn, *The Essential Tension: Selected Studies of Scientific Tradition and Change* [Chicago: The University of Chicago Press, 1977], pp. xi–xii).

27 F. J. J. Buytendijk, *Het Kennen van de Innerlijkheid.*

28 D. J. van Lennep, *Gewogen, Bekeken, Ontmoet in het Psychologisch Onderzoek* (inaugural lecture at Utrecht) (The Hague: Mart. Nijhoff, 1949).

29 D. J. van Lennep, *Gewogen, Bekeken, Ontmoet,* p. 10.

30 That was the case, at least, in Van Lennep's inaugural lecture and in D. J. van Lennep, *Psychotechniek als Kompas voor het Beroep.* As told in Chapter 2, in the second half of the 1950s Van Lennep gradually changed his views on this. I will also come back to this in the present chapter.

31 M. J. Langeveld, *Handelen en Denken in de Opvoeding en de Opvoedingswetenschap* (inaugural lecture at Amsterdam) (Groningen: J. B. Wolters, 1942), p. 9.

32 M. J. Langeveld, *Over het Wezen,* p. 32.

33 M. J. Langeveld, *Over het Wezen,* p. 33.

34 For a more extensive English-language description of Scheler's thought, see H. Spiegelberg, *The Phenomenological Movement,* chapter V.

35 Spiegelberg published some letters from Husserl to a student, E. P. Welch, in which Husserl makes clear statements on Schelerian philosophy: "The fact that someone was my academic student or has become a philosopher under the influence of my writings does not therefore mean by far that he has penetrated to a real understanding of the inner meaning of my, the original phenomenology and its method. . . . This is true of . . . even such famous men as Max Scheler . . . , in whose philosophies I see merely ingenious relapses into the old philosophical naivetes" (H. Spiegelberg, *The Context of the Phenomenological Movement* [The Hague: M. Nijhoff, 1981], p. 181).

36 Translation: M. Scheler, *Formalism in Ethics and Non-formal Ethics of Value* (transl. M. S. Frings and R. L. Funk) (Evanston, IL: North Western University Press, 1973).

37 "Ein Zusammenhang in dem die moralische und die theoretische Welt aneinander – wie mit Klammern – ewig gebunden sind." M. Scheler, *Vom Ewigen im Menschen: Gesammelte Werke* 5 (Bern: Francke Verlag, 1921; repr. 1974) p. 90–1.

38 Included in M. Scheler, *Vom Ewigen im Menschen.*

39 M. Scheler, *Vom Ewigen im Menschen,* p. 90.

40 W. J. M. Dekkers, *Het Bezielde Lichaam;* R. H. J. ter Meulen, *Ziel en Zaligheid.*

41 R. H. J. ter Meulen, *Ziel en Zaligheid,* p. 184. In his study of the reception of psychology and psychoanalysis among Dutch Catholics, Ter Meulen reaches conclusions with regard to the personalist nature of Catholic phenomenology that converge strongly with the conclusions drawn in this chapter regarding the phenomenology of the Utrecht School. In his preface Ter Meulen mentions that he had to adjust his original plan to study the relation of phenome-

nology, Christendom (particularly Catholicism), and the social sciences in the Netherlands: "Phenomenology turned out to have had by no means the influence on the practice of the social sciences among Catholics in the Netherlands as was assumed in the formulation of the project proposal. The reception and further development of psychology and psychiatry among the Catholics turned out to be rather dominated by the discussion between naturalism and Christian personalism and by the ancient scholastic discussion between intellectualism and voluntarism" (p. 7).

42 See Chapter 4.

43 See Ph. A. Kohnstamm, *Hoe mijn Bijbels Personalisme Ontstond* (Haarlem: Tjeenk Willink, 1934).

44 Already at this time the difference between the Amsterdam psychologists and the personalistic thinkers was becoming obvious. From Kohnstamm's Amsterdam period comes an anecdote that relates how Révész, making fun of the personalistic preoccupations with values, could deeply wound Kohnstamm with the claim: "Ich bin ein besserer Mensch wie Sie" ("I am a better human than you") (De Groot, told to author).

45 The sources for these facts on the relationship between Langeveld and Kohnstamm are Langeveld's curriculum vitae in the "Archief Van Dael," included in the Archives of the History of Dutch Psychology, Groningen, as well as interviews with Langeveld held in 1986 and 1987.

46 Langeveld (1986), interview.

47 According to W. Wijga, who was one of the staff members (and from 1952 the director) of the Foundation. In the foreword of his dissertation Van Lennep also recalls the wartime contacts with Plessner (D. J. van Lennep, *Psychologie van de Projectieverschijnselen*).

48 This book has been translated into many languages, including German, Spanish, Portuguese, and Italian. The English-language version is: F. J. J. Buytendijk, *Woman: A Contemporary View*. The original Dutch version is *De vrouw, haar Natuur, Verschijning en Bestaan: Een Existentieel-psychologische Studie* (Utrecht: Het Spectrum, 1951).

49 Translation: S. de Beauvoir, *The Second Sex* (transl. H. M. Parshley) (London: Johnathan Cape, 1953).

50 Buytendijk, *De Vrouw*, reprint 1958, p. 27.

51 Buytendijk, *De Vrouw*, reprint 1958, p. 26.

52 More can be found on the relationship between Buytendijk and Merleau-Ponty in C. E. M. Struyker Boudier, "Buytendijk en Merleau-Ponty, Relaas van een Relatie," *Algemeen Nederlands Tijdschrift voor de Wijsbegeerte* 76 (1983):228–46.

53 Ter Meulen wrote about this: "In his later work Buytendijk made much use of the existentialist terminology of Merleau-Ponty. There was however no question of a fundamental change in Buytendijk's theoretical position" (R. H. J. ter Meulen, "De Deugd van een Geboren Heer," p. 10).

54 M. J. Langeveld (1987), interview.

55 W. Banning (1938), quoted in J. Lindeboom, *Geschiedenis van de Barchembeweging, MCMVIII–MCMLVIII* (Amsterdam: Barchembeweging, 1958), p. 193.

56 M. J. Langeveld (1987), interview.

57 M. de Keizer, *De Gijzelaars van Sint Michielsgestel: Een Elite-beraad in Oorlogstijd* (Alphen aan den Rijn: Sijthoff, 1979).

58 Bank has written a history of the NVB from which a number of the facts mentioned here are derived (J. Bank, *Opkomst en Ondergang van de Nederlandse Volksbeweging (NVB)* [Deventer: Kluwer, 1978]).

59 On Kruijt and the way in which he expressed his personalistic socialist ideals in his cultural sociology, see M. Gastelaars, *Een Geregeld Leven: Sociologie en Sociale Politiek in Nederland 1925–1968* (Amsterdam: SUA, 1985).

60 Ph. A. Kohnstamm, W. P. J. Pompe, and H. R. Hoetink, *Personalistisch Socialisme naar drieërlei fundering* (Amsterdam: Nederlandse Volksbeweging, 1945).

61 The Purging Decree (*Zuiveringsbesluit*) concerning persons in governmental service stated that officials could be fired (*geschorst*) or suspended (*gestaakt*). The former measure was taken against Buytendijk's predecessor Roels. When one was fired, one got only a low benefit. Langeveld was suspended, which meant that he kept his salary but had to stop working.

62 Source: State Archives Utrecht, Archives Board of Governors, University of Utrecht, nr. 2914.

63 Source: State Archives Utrecht, Archives Board of Governors, University of Utrecht, nr. 2914. A structural problem of personalistic socialism also plays a role here. Personalistic socialists did occasionally see their own ideals realized by National Socialism. De Keizer (*De Gijzelaars van Sint Michielsgestel*) goes extensively into the ambivalent relations of the French personalist Mounier with National Socialism. Mounier worked with the collaborating Vichy regime on training a new "combative vanguard for the future" in the so-called Ecole des Cadres. After he withdrew from this group, he joined the resistance but still had difficulties after the war. Also known is the example of the Belgian socialist De Man who after the war was condemned for his support of National Socialism (see P. Dodge, *A Documentary Study of Hendrik de Man: Socialist Critic of Marxism* [Princeton: Princeton University Press, 1979]; D. Pels, "Hendrik de Man and the Fascist Temptation," *History of the Human Sciences* 6 [1993]:75–97). In the Netherlands there was also a close connection between the NVB and the *Nederlandse Unie,* an organization that in some respects collaborated with the occupying forces (see also note 66). It must be noted, however, that in 1932 Banning had already written a warning leaflet on National Socialism in the series of publications of the Workers' Community, and that shortly afterward he became chairman of the antifascist (and anticommunist) movement *Eenheid door Democratie* (Unity by Democracy). Buytendijk also opposed National Socialism in the 1930s on the basis of his personalistic thought (F. J. J. Buytendijk, "Rassenwaan en Medische Wetenschap," in *Het Christendom Bedreigd (door Rassenwaan en Jodenhaat)* [Amsterdam, Fidelitas, 1936], pp. 28–36).

64 This was related to me by the Groningen professor of psychology J. Th. Snijders, who himself became a member of the NVB.

65 A year later Langeveld was informed by the NVB leadership that this text could not be published because of a lack of funds (source: International Institute for Social History Amsterdam, Archives NVB).

66 This was also due to the participation of the *Nederlandse Unie* (the Dutch Union) in NVB activities. The Unie members, De Quay (also professor of psychology at Tilburg) and Einthoven, were also hostages in the Sint Michielsgestel Camp and belonged to the NVB leadership. At the beginning of the war, the Unie had already been proposing a sort of Dutch socialism with an antiparliamentary and in certain respects even a pro-German tendency. The latter especially meant that after the war the Unie members were the subject of controversy. The close relations of the NVB with the Nederlandse Unie impeded the NVB's plans (J. Bank, *Opkomst en Ondergang van de Nederlandse Volksbeweging (NVB)*; M. de Keizer, *De Gijzelaars van Sint Michielsgestel*).

67 See, for example, the issue on Roman Catholicism in the Protestant journal *Wending* (January 1948), in which the authors express very profound suspicions of "Papism."

68 A. L. T. Notten and B. van Gent, "Geschiedenis," in B. van Gent et al., *Welzijnswerk en Wetenschap* (Alphen aan den Rijn: Samsom, 1984), pp. 109–25.

69 W. Banning, *De Dag van Morgen: Schets van een Personalistisch Socialisme: Richtpunt voor Vernieuwing van ons Volksleven* (Amsterdam: Ploegsma, 1945).

70 W. Banning, *De Dag van Morgen*, pp. 138–9.

71 M. J. Langeveld, "De Opvoeding van de Student aan de Universiteit," in M. J. Langeveld, *Verkenning en Verdieping* (Purmerend: J. Muusses, 1950), pp. 322–36 (p. 332).

72 M. J. Langeveld, *Verkenning en Verdieping*, p. 336.

73 M. J. Langeveld (1948) at a conference of the International Association of University Professors and Lecturers, included in M. J. Langeveld, *Verkenning en Verdieping*, pp. 314–22 (p. 317).

74 Langeveld, who was from a Baptist family, regarded himself as an unaffiliated Christian; Buytendijk was originally Protestant, became strictly Presbyterian, and then devoutly Catholic; Van Lennep was Protestant, but converted to Catholicism in adulthood; Dijkhuis was a cradle Catholic, but not orthodox; and Kouwer was an atheist.

75 Buytendijk's name does not appear on the list of signatories to the NVB manifesto. It is said of Buytendijk that he never really committed himself to a political party or movement, but he does appear to have had sympathies with the Labor Party (PvdA).

76 F. J. J. Buytendijk, *Erziehung zur Demut* (Leipzig: Neue Geist Verlag, 1928).

77 M. J. Langeveld (1987), interview.

78 F. J. J. Buytendijk, "De Zending der Wetenschap," in G. Marcel et al., *Katholieke Synthese* (Nijmegen/Utrecht: Dekker and Van de Vegt, 1939), pp. 109–39.

79 F. J. J. Buytendijk, *De Idealen van den Idealen Student* (Utrecht: Het Spectrum, 1945), p. 28–9.

80 It is common practice in Dutch universities to consult fellow faculties in other cities before appointing a professor. On the original list, drawn up on the

recommendations of professors of psychology throughout the country, De Groot got the highest score by being named four times. Two others were named three times and five others, including Van Lennep (chosen by Rümke), were nominated once. Source: State Archives Utrecht, Archives of the Board of Governors, University of Utrecht, nr. 711.

81 G. van der Leeuw, *Balans van Nederland* (Amsterdam: H. J. Paris, 1945). Van der Leeuw was also the author of *Einführung in die Phenomenologie der Religion* (München: Ernst Reinhardt, 1925), which was translated into many languages.

82 During the occupation, in September 1941, by decree of the secretary general of the Department of Education, a separate regulation for exams in psychology was introduced into the Academic Statutes, after the German model. Like all other decrees made during the occupation it was withdrawn after liberation, but by Royal Decree of 21 September 1946 it was again included, unaltered, in the Academic Statutes.

83 Source: the (unpublished) minutes of the Sassen Commission (Archives of the History of Dutch Psychology, Groningen), and a discussion between De Groot and Kohnstamm. See Ph. A. Kohnstamm, "Paedagogiek en Psychologie," *Paedagogische Studieën* 26 (1949):378–86; A. D. de Groot, "Psychologie en Paedagogiek: Een Wederwoord aan Prof. Kohnstamm," *Paedagogische Studieën* XXVII (1950):5–74.

84 H. C. J. Duijker, "De Opleiding der Psychologen," *Nederlands Tijdschrift voor de Psychologie* 4 (1949):341–67.

85 A. D. de Groot, *De Psycholoog in de Maatschappij* (inaugural lecture at Amsterdam) (Amsterdam: Noord Hollandse Uitgevers Maatschappij, 1949), p. 18.

86 Typical of the differences in this respect is the small discussion that had taken place in the last meeting of the Sassen Commission. Willems, who had said hardly anything, had then suddenly remarked that *statistics* was missing from the program that had been formulated. Langeveld, however, replied that it was not feasible to mention "all aspects of psychology and its supporting sciences." To this Sassen added that at that stage only editorial changes were still possible and then closed the meeting (source: Minutes of the 22nd meeting of the Sassen Commission, Archives of the History of Dutch Psychology, Groningen). Nevertheless, statistics was one of the seven *optional* subjects in the commission's final proposal.

87 A. D. de Groot, *De Psycholoog in de Maatschappij*, p. 17. He also added to the latter that it is somewhat exaggerated.

88 Various themes emerge from the correspondence between Rutten and the Board of Governors of the University and between the Board and the Humanities Faculty (State Archives Utrecht, Archives of the Board of Governors, University of Utrecht, nrs. 2429 and 2495).

89 In Dutch the poem reads: "Helpt Heer, nu ook de Buytendijken breken/ Van Nijmegen tot aan het oude Sticht/ alom de leer der vaderen wordt ontwricht/ en eerbied uit de wereld is geweken." *Parasol* was a student magazine founded in 1953 in opposition to the university organ *Sol Iustitiae Illustra Nos*. *Parasol* caused a furor by accusing Buytendijk of plagiarism in his book

De Vrouw (for more on this affair, see A. Fransen, "Parasol, een korte voorzomer in de ontstaansgeschiedenis van Tirade," *Tirade* 29 [1985]:517–46).

90 D. J. van Lennep, *Psychotechniek als Kompas voor het Beroep* (Utrecht: De Haan, 1949); D. J. van Lennep, "De Ontwikkeling van het Testonderzoek in de Bedrijfspsychologie," *Nederlands Tijdschrift voor de Psychologie* 12 (1957):270–7.

91 D. J. van Lennep, *Psychotechniek als Kompas voor het Beroep,* pp. 271 and 276. This did not mean though that the old intuitive methods would be abandoned in the Foundation for Psychotechniks (see Chapter 2).

92 B. J. Kouwer, "Moderne Magie" (1953), in B. J. Kouwer, *Persoon en Existentie* (Groningen: Wolters Noordhoff, 1977), p. 39.

93 B. J. Kouwer, *Gewetensproblemen van de Toegepaste Psychologie* (inaugural lecture at Groningen, 1955), pp. 12–13.

94 With regard to Van Lennep there is some obscurity here. J. H. Dijkhuis (1988, interview) recalled that after the lecture he drove back to Utrecht with Van Lennep and that Van Lennep was "livid." Others (W. Wijga, J. Bremer, as related to the author) who also knew Van Lennep well, however, doubt that there was a rupture between the two men.

95 Here it is a matter of the idea Buytendijk expressed in many ways: The church produces too many neurotic believers. People must be given more freedom so that they can develop a justified, healthy faith. This is very clearly seen in F. J. J. Buytendijk, *Gezondheid en Vrijheid* (lecture given on the occasion of the fortieth anniversary of the RK Limburgse Groene Kruis) (Utrecht: Katholieke Centrale Vereniging voor de Geestelijke Volksgezondheid, 1950).

96 B. J. Kouwer, *Gewetensproblemen van de Toegepaste Psychologie,* p. 9.

97 J. Linschoten, *Op Weg naar een Fenomenologische Psychologie* (Utrecht: Bijleveld, 1959). Translation into English: J. Linschoten, *On the Way Toward a Phenomenological Psychology.*

98 See J. Linschoten, "Die Unumgänglichkeit der Phänomenologie," in V. E. von Gebsattel, P. Christian, and W. J. Revers, *Jahrbuch für Psychologie, Psychotherapie und Medizinische Anthropologie* (Freiburg und München: Verlag Karl Alber, 1962), pp. 177–85; J. Linschoten, "Fenomenologie en Psychologie," *Algemeen Nederlands Tijdschrift voor de Wijsbegeerte* 55 (1963), pp. 113–22. For an English-language description of the development of Linschoten's thought, see also S. J. S. Terwee, "The Case of Johannes Linschoten's Apostasy: Phenomenological versus Empirical-Analytical Psychology," in *Hermeneutics in Psychology and Psychoanalysis* (Heidelberg and New York: Springer, 1990), chapter 3.

99 This was known only in closed circles. It was kept secret from the outside world, as shown, for example, by the obituaries on Linschoten. They expressed their disappointment only among themselves; see, for example, the letters quoted by Ter Meulen, in the archives of the Catholic Documentation Center Nijmegen (R. H. J. ter Meulen, *Ziel en Zaligheid,* pp. 355–6).

100 This position was taken by R. H. J. ter Meulen, *Ziel en Zaligheid* (pp. 218 and 355–6).

101 F. J. J. Buytendijk, *De Vrouw,* p. 27.

102 These developments are shown by data from surveys by C. Middendorp, *Ontzuiling, Politisering en Restauratie in Nederland: De Jaren 60 en 70* (Meppel: Boom, 1979). Brinkgreve and Korzec analyzed the contents of the problem page from the women's magazine *Margriet* and concluded that in the second half of the 1960s the period when people were given general rules for proper behavior came to an end. Instead, the norm for behavior was set by the individual conscience (C. Brinkgreve and M. Korzec, *Margriet Weet Raad: Gevoel, Gedrag, Moraal in Nederland, 1938–1978*. Utrecht: Het Spectrum, 1978).

103 For an analysis of the mechanisms with which modern empirical psychology, for example, "obliges people to be free," see N. Rose, *Governing the Soul: The Shaping of the Private Self* (London and New York: Routledge, 1989; repr. 1991).

104 On the disappearance of personalistic socialist values in PvdA first with the rise of the New Left and somewhat later with the return to Marxism, see A. Bleich, *Een Partij in de Tijd; Veertig Jaar Partij van de Arbeid 1946–1986* (Amsterdam: De Arbeiderspers, 1986); D. Pels, "Willem Banning: Voor en Tegen," in M. Krop et al., eds. *Het Negende Jaarboek voor het Democratisch Socialisme* (Amsterdam: De Arbeiderspers/ Wiardi Beckman Stichting, 1988), pp. 134–70.

105 L. Boon and P. A. Vroon, "Psychologie," in H. W. von der Dunk, et al., eds., *Tussen Ivoren Toren en Grootbedrijf: De Utrechtse Universiteit, 1936–1986* (Maarssen: Schwartz, 1986) pp. 483–6 (p. 485); respectively D. Pels, "Willem Banning: Voor en Tegen," p. 168.

106 M. J. Langeveld, *Over het Wezen der Paedagogische Psychologie en de Verhouding der Psychologie tot de Paedagogiek* (Groningen: J. B. Wolters, 1951), p. 27.

107 For example: B. J. Schreurs, "Psychologie, Filosofie en de Klamme Handjes," *Wijsgerig Perspectief* 12 (1971–2):336–61. From the end of the 1960s to the early 1980s, students and younger academics at other universities did argue for a "critical psychology." Apart from Marxist writings in general, people were chiefly inspired, as I related Chapter 1, by the Berlin psychologist Klaus Holzkamp; another example is the book by P. van den Dool and A. Verbij, eds., *Van Nature Maatschappelijk: Overzicht van de Kritische Psychologie* (Amsterdam: Sua, 1981). However, "critical psychology" never became a very powerful movement in the Netherlands. The dominant methodology remained that of empirical-analytical thinking. The Marxist-inspired journal *Psychologie & Maatschappij*, founded in 1977, is now flourishing as a journal of theoretical psychology, history of psychology, and general reflection on psychology.

108 These letters are to be found in the Catholic Documentation Center Nijmegen, Buytendijk Archives, nr. 397.

Chapter 4. The neurotic paradox of clinical psychology

1 J. J. Groen (1987), interview.

2 J. T. Barendregt, "Moge de Crisis Blijven" (1977), in *De Zielenmarkt: Over Psychotherapie in Alle Ernst* (Meppel: Boom, 1982), pp. 145–9 (pp. 146–7).

3 This formulation is from the psychologist N. H. Frijda, who ascribes it to opponents of empirical-analytical methodology. The context of the quote will be discussed later in this chapter.

4 See L. J. Pongratz, *Lehrbuch der Klinischen Psychologie* (Göttingen: Hogrefe, 1963), pp. 20–34.

5 I elaborated upon this in T. Dehue, "The 'Label Criterion' in the Historiography of Scientific Disciplines: The Case of Clinical Psychology in the Netherlands." In A. Pankalla and R. Stachowsky, eds., *Studies in the History of Psychology and the Social Sciences, 5* [Proceedings 12th Annual Meeting of Cheiron Europe, Poznan, 1993]. Poznan: The Adam Mickiewicz University (forthcoming). Comparable issues are raised in R. Smith, "Does the History of Psychology Have a Subject?" *History of the Human Sciences* 1 (1988):147–79 and R. de Wilde, *Discipline en Legende: De Identiteit van de Sociologie in Duitsland en de Verenigde Staten 1870–1930* (Amsterdam: Van Gennep, 1992). Many complicated issues are connected to the topic of scientific discipline formation in general or the formation of psychology as an independent discipline. See, for instance, W. R. Woodward, "World View and Scientific Discipline Formation: How East German Science Studies Contributed to the Fall of the Cultural Wall," in W. R. Woodward and R. S. Cohen, eds., *World View and Scientific Discipline Formation* (Dordrecht: Kluwer, 1991), pp. 1–17.

6 J. E. van Lennep (1987), interview. Dr. J. E. van Lennep should not be confused with the Utrecht Professor D. J. van Lennep discussed earlier in this book.

7 On early clinical psychology in the United States see, for example, J. R. Reisman, *A History of Clinical Psychology* (New York: Hemisphere Publishing Corporation, 1991); B. Richards, "Lightner Witmer and the Project of Psychotechnology," *History of the Human Sciences* 1 (1988):201–21.

8 According to the founder of the Association, J. J. Dijkhuis, in an unpublished letter to the NIPP, dated 5 April 1954. Archives of the History of Dutch Psychology, Groningen.

9 However, in the United States intelligence tests were first used in a medical context (L. Zenderland, "The Debate over Diagnosis: Henry Goddard and the Medical Acceptance of Intelligence Testing," in M. Sokal, ed., *Testing and American Society, 1890–1930* (New Brunswick and London: Rutgers University Press, 1987), pp. 46–75.

10 For a more extensive explanation of anthropological medicine and for more references see W. J. M. Dekkers, *Het Bezielde Lichaam: Ontwerp van een Antropologische Fysiologie en Geneeskunde Volgens F. J. J. Buytendijk* (Zeist: Kerkebosch, 1985). For the influence of anthropological thought in Dutch psychiatry and geriatrics, see A. A. M. Prins, "The Case of the Untraceable Alzheimer Patients: Medical Practices and Intellectual Traditions of Psychiatrists and Geriatrists in the Netherlands 1900–1988," in I. Löwy et al., eds., *Medicine and Change: Historical and Sociological Studies of Medical Innovation* (London: John Libbey, 1993), pp. 395–421; see also D. Wyss, "Die Antropologisch Existentielle Psychologie und ihre Auswirkungen Insbesondere auf die Psychiatrie und Psychotherapie," in H. Balmer, ed., *Die Psychologie im 20. Jahrhundert, Band I* (Zürich: Kindler, 1976), pp. 461–9.

11 For the background and considerations in the founding of the Free University

and the Valerius Clinic, see M. van Os and W. J. Wieringa, eds., *Wetenschap en Rekenschap, 1880–1980: Een Eeuw Wetenschapsbeoefening en Wetenschapsbeschouwing aan de Vrije Universiteit* (Kampen: Kok, 1980). And also Vereniging tot Christelijke Verzorging van Geestes- en Zenuwzieken in Nederland, *Een Halve Eeuw Arbeid op Psychiatrisch-neurologisch Terrein, 1910–1960* (Wageningen: Zomer & Keuning, 1960). The quotes in this section are on page 186 of the first mentioned work, and on page 16 of the second.

12 Two exceptions to this were the psychologist A. A. Grünbaum who had studied in Würzburg, among other places, and the psychologist J. van Essen, who had graduated in psychology at Vienna. Both were employed in the Valerius Clinic, too.

13 L. van der Horst, "Geneeskundige Behandeling en Geestelijke Verzorging," *Nederlands Tijdschrift voor Psychologie VII* (1939):35–47 (p. 36).

14 However, in the daily practice of this psychiatry the application of this doctrine's accusatory element seems to have been minimal. See J. A. van Belzen, *Psychopathologie en Religie: Ideeën, Behandeling en Verzorging in de Gereformeerde Psychiatrie, 1880–1940* (Kampen: Kok, 1989).

15 R. Abma, *Methodisch zonder Confessie* (Nijmegen: Psychologisch Laboratorium KUN, 1983) p. 66.

16 See H. Struyker Boudier, "Psychosomatiek en de Antropologische Geneeskunde in het Werk van J. J. G. Prick," in K. J. M. van de Loo, ed., *Psychosomatiek: Theoretische en Klinische Bijdragen* (Baarn: Ambo, 1986):36–60.

17 J. J. G. Prick, "Over de Psychologische, Psychopathologische en Personale Aspecten van de Psychosomatologie," in J. Groen, J. J. G. Prick, and H. Faber, eds., *Psychosomatiek, Geneeskunde en Mensbeschouwing* (Amsterdam: Scheltema en Holkema, 1953) pp. 42–3.

18 R. Abma, *Methodisch zonder Confessie*, p. 68.

19 K. J. M. van de Loo, *De Klinische Psychologie in Dienst van de Problematiek van de Essentiële Hypertensie: Een Psychologische Bijdrage tot de Psychosomatische Geneeskunde* (Nijmegen: Dekker & Van de Vegt, 1952), p. 224.

20 Known in the Netherlands at that time as "lector."

21 This was the Sint Willibrordus Stichting in Heiloo.

22 J. J. Diamant et al., *Veertig Jaar Vakgroep Psychologen* (unpublished report on the occasion of the fortieth anniversary of the psychology department of the Sint Willibrordus Stichting in Heiloo, 1987), Archives of the History of Dutch Psychology, Groningen.

23 The staff of the Provinciaal Ziekenhuis in Santpoort, for example, came to Heiloo to see what a psychologist could do, and in 1948 it appointed the psychologist P. Boeke. Also in 1948 the Nijmegen psychologist E. Breukers became Rümke's assistant at the Psychiatrische Universiteitskliniek of the Utrechtse Academisch Ziekenhuis. J. E. van Lennep was appointed in both the Valeriuskliniek and at the University of Amsterdam (J. E. van Lennep, 1987, interview).

24 J. J. Dijkhuis in an unpublished letter to the NIPP, dated 5 April 1954, Archives of the History of Dutch Psychology, Groningen.

25 It is true that the first cooperation between a psychiatrist and a psychologist in the Netherlands was the one between Heymans and Wiersma, which began around the turn of the century (see L. K. A. Eisenga, *Geschiedenis van de Nederlandse Psychologie* [Deventer: Van Loghum Slaterus, 1978], p. 50), but Heymans and Wiersma had no significant influence on either the genesis of the field of clinical psychology or on the work of the first Dutch clinical psychologists.

26 In Germany, where the Dutch found the philosophical inspiration for their anthropological medicine, there was, as far as I know, no connection between such medicine and the first group of psychologists to call themselves clinical psychologists. In his history of German clinical psychology Rechtien mentions that until the 1970s the field had no involvement at all in somatic and psychosomatic illnesses. The first German clinical psychologists clearly expressed the importance of cooperation with doctors, in definitions of their professional field, but they concentrated on the relationship between the doctor and the patient, the relation of the patient to his sickness, and so on (W. Rechtien, "Geschichte der Klinischen Psychologie," in H. Lück, H. Grünwald, U. Geuter, R. Miller, W. Rechtien, eds., *Sozialgeschichte der Psychologie* [Opladen: Leske und Budrich, 1987], pp. 209–25. Nor do Hörmann and Nestmann mention any connection between the first clinical psychologists in Germany and anthropological medicine and psychosomatics (G. Hörmann and F. Nestmann, "Die Professionalisierung der Klinischen Psychologie und die Entwicklung neuer Berufsfelder in Beratung, Sozialarbeit und Therapie," in M. Ash and U. Geuter, eds., Geschichte der Deutschen Psychologie im 20. Jahrhundert [West Deutscher Verlag, 1985], pp. 252–85). In England before 1940 some psychologists were working in a psychiatric context, but there, too, it was not against the background of a Christian anthropological psychiatry. The same is true of the postwar growth of psychology in a psychiatric context, which was, as in the United States, the direct result of the application of psychology in the military (L. S. Hearnshaw, *A Short History of British Psychology, 1840–1940* [London: Methuen, 1964]; N. Rose, *The Psychological Complex* [London: Routledge, 1985]. Neither does the history of British psychology in the postwar period mention such a relationship (N. Rose, *Governing the Soul* [London: Routledge, 1989; repr. 1991]; J. S. Reisman, *A History of Clinical Psychology*), and no trace of this relation was found in G. Zilboorg, *A History of Medical Psychology* (New York: Norton, 1949; repr. 1969).

27 B. Richards, "Lightner Witmer and the Project of Psychotechnology"; J. M. Reisman, *A History of Clinical Psychology*.

28 G. Kraus, "De Samenwerking van Psychiater en Psycholóog," *Nederlands Tijdschrift voor de Psychologie* 11 (1956):221–9 (p. 225).

29 J. J. G. Prick, "De Verhouding van Psychologie en Psychopathologie," *Nederlands Tijdschrift voor de Psychologie* 2 (1947):3–20 (p. 4).

30 L. van der Horst, "Inleiding," *Nederlands Tijdschrift voor de Psychologie* 1 (1957):180–1 (p. 180).

31 K. J. M. van de Loo, "Over de Samenwerking tussen Psychiater en Psycholoog, in Het Bijzonder de Klinisch Psycholoog," *Nederlands Tijdschrift voor de Psychologie* 11 (1956):183–205 (p. 188).

32 L. van der Horst, "Over de Methodiek der Psychologie," *Nederlands Tijdschrift voor Psychologie* I (1933):40–53 (p. 52).

33 K. J. M. van de Loo, *De Klinische Psychologie in Dienst van de Problematiek van de Essentiële Hypertensie: Een Psychologische Bijdrage tot de Geneeskunde* (thesis) (Nijmegen: Dekker & Van de Vegt, 1952):224.

34 K. J. M. van de Loo, *De Klinische Psychologie*, p. 224.

35 B. J. Kouwer, *Tests in de Psychologische Practijk* (Utrecht: Bijleveld, 1952), pp. 22 and 33.

36 J. J. Groen (1987), interview.

37 S. J. Vles and J. J. Groen, "Een Onderzoek van de Persoonlijkheidsstructuur van Jeugdige Asthma-patienten met behulp van de Behn-Rorschachtest," *Nederlands Tijdschrift voor de Psychologie* 6 (1951):29–73.

38 J. J. Groen (1987), interview.

39 De Groot gave a more detailed report of the steps taken for this than Barendregt did himself. See A. D. de Groot, *Methodologie*. English edition: *Methodology* (The Hague: Mouton, 1969):139–50.

40 H. J. Gibson, *Hans Eysenck: The Man and His Work* (London: Peter Owen, 1981).

41 H. J. Eysenck, *The Scientific Study of Personality* (London: Routledge and Kegan Paul, 1952), p. 18.

42 J. T. Barendregt, "Is de Waarde van Psychotherapie Bewijsbaar?" (1957), in *De Zielenmarkt,* pp. 95–105 (p. 99–100).

43 H. A. van der Sterren, "Discussion Remarks" (unpublished, 1961, private archives of J. Bastiaans, Leiden).

44 J. Bastiaans, J. T. Barendregt, and A. W. Vermeul-Van Mullem, "Antwoord van de Onderzoekers op de Binnengekomen Kritieken" (unpublished, 1961, private archives of J. Bastiaans, Leiden).

45 J. Bastiaans (1988), interview.

46 J. T. Barendregt, *Research in Psychodiagnostics* (The Hague: Mouton, 1961).

47 J. Bendien, "Een Psychologie Die Zo Wetenschappelijk Wil Zijn Dat Het Niet Wetenschappelijk Meer Is," *Nederlands Tijdschrift voor de Psychologie* 12 (1959):543–57 (p. 549).

48 N. H. Frijda, "Om Derwille van het Psychologisch Experiment," *Nederlands Tijdschrift voor de psychologie* 16 (1961):110–17 (p. 110).

49 J. T. Barendregt, "Antwoord op 'Enkele Opmerkingen naar Aanleiding van Barendregts Studie over de Waarde van Klinische Predictie,' " *Nederlands Tijdschrift voor de Psychologie* 14 (1959):335–6; J. T. Barendregt, "Een Criticus Die Zo Negatief Wil Zijn, Dat Hij Positief Is," *Nederlands Tijdschrift voor de Psychologie* 17 (1962):558–61.

50 It actually took some time and effort for the NIPP and the Association to agree on the conditions with regard to this joining the Association. See T. A. Veldkamp and P. van Drunen, *Psychologie als Professie: 50 Jaar Nederlands Instituut van Psychologen* (Assen: Van Gorcum, 1988), p. 20–1.

51 Figures mentioned in a letter from the group's secretary J. R. M. van Ratingen

on 28 October 1957 (unpublished, Archives of the History of Dutch Psychology, Groningen).

52 K. H. Bouman (1939) quoted in T. van der Grinten, *De Vorming van de Ambulante Geestelijke Gezondheidszorg: Een Historisch Beleidsonderzoek* (Baarn: Ambo, 1987), p. 78.

53 This observation is derived from P. van Lieshout, "Veertig Jaar Geestelijke Volksgezondheid: Een Analyse van het MGv," *Maandblad voor Geestelijke Volksgezondheid* 40 (1985):1243–73. See also A. Mol and P. van Lieshout, *Ziek Is Het Woord Niet* (Nijmegen: SUN, 1989).

54 For example, F. J. J. Buytendijk, *Gezondheid en Vrijheid. Rede Gehouden Ter Gelegenheid van het 40-jarig Bestaan van het RK Limburgse Groene Kruis* (Utrecht: Katholieke Centrale Vereniging voor de Geestelijke Volksgezondheid, 1950).

55 See, for example, J. G. Fernhout, *Psychotherapeutische Zielzorg* (Baarn: Bosch & Keuning, 1950); C. W. du Boeuff and P. C. Kuiper, *Psychotherapie en Zielzorg* (Utrecht: Bijleveld, 1950).

56 H. R. Wijngaarden, *Nog Enige Beschouwingen over Conflictuologie* (public lecture) (Amsterdam: Vrije Universiteit, 1978), p. 3.

57 J. T. Barendregt, *Psychologische Persoonlijkheidsleer* (inaugural lecture, University of Amsterdam) (Amsterdam: Polak & Van Gennep, 1962).

58 A. D. de Groot, "Scientific Personality Diagnosis," *Acta Psychologica* 10 (1954):220–41.

59 J. T. Barendregt, "Jaarverslag met Voorgeschiedenis," *Bulletin Persoonlijkheidsleer Universiteit van Amsterdam* I (1972).

60 This name was derived from the psychologist Monte Shapiro, who from the London Maudsley Clinic gave reports on his experiences with the experimental study of the specific factors that play a role in the various complaints of individual patients.

61 I. van Krogten (1987), interview.

62 Barendregt's assistant Hamers wrote a thesis on the EID and its failure. The figures mentioned are given in this document (H. Hamers, *De Experimentele Individuele Diagnostiek* [Department of Personality Psychology, University of Amsterdam, 1967]).

63 J. T. Barendregt, "Klinische Psychologie Is Meer Dan Therapie Bedrijven," *De Psycholoog* 9 (1974):127–36 (pp. 130–1).

64 J. T. Barendregt, "Jaarverslag met Voorgeschiedenis."

65 For a history of behavioral therapy, see A. Schorr, *Die Verhaltenstherapie* (Weinheim und Bazel: Beltz, 1984).

66 J. T. Barendregt, "Comments to Eysenck's 'The Effects of Psychotherapy,' " *International Journal of Psychiatry* 1 (1965):161–3.

67 Whereas 69 first-year students were registered in the Psychology Faculty of the University of Amsterdam in 1960, in 1965 there were already 224 and in 1970 there were 405 (H. de Haan, "Statistisch Jaaroverzicht," Universiteit van Amsterdam: Subfaculteit Psychologie, 1982). From the very beginning the Department of Personality Psychology (i.e., clinical psychology) was the

one that attracted the most students among psychology departments of the University of Amsterdam.

68 G. J. S. Wilde, "Psychologie in de Psychiatrische Universiteitskliniek," *Mededelingen van het Instituut voor Clinische en Industriële Psychologie* (1964), p. 12.

69 E. Raven, "Vraagstellingen bij het Psychologisch Onderzoek in de Tweede Psychiatrische Universiteitskliniek te Utrecht," *Mededelingen van het Instituut voor Clinische en Industriële Psychologie* (1966), p. 10.

70 W. T. A. M. Everaerd, "Vraagstelling en Methode in het Individueel Psychologisch onderzoek," *Mededelingen van het Instituut voor Clinische en Industriële Psychologie* 6 (1965):6–8. The example given is described by the author as one derived from the British psychologist David Payne.

71 F. Beenen, "Klinische Psychologie en Besluitvorming in de Psychiatrische Kliniek," in A. P. Cassee et al., eds., *Klinische Psychologie in Nederland 1* (Deventer: Van Loghum Slaterus, 1973):29–56. See also P. E. Boeke, *Psychologie in de Medische Situatie* (public lecture, RU Groningen) Groningen: J. B. Wolters, 1965).

72 J. J. Dijkhuis, *Veertig Jaar en Verder?* (lecture given on the fortieth anniversary of the psychological department of the Sint Willibrordus Stichting in Heiloo in 1987) (unpublished, Archives of the History of Dutch Psychology, Groningen); J. J. Dijkhuis (1988), interview with P. van Drunen and M. Wopereis (transcript: Archives of the History of Dutch Psychology, Groningen).

73 C. Brinkgreve, *Psychoanalyse in Nederland* (Amsterdam: De Arbeiderspers, 1984), p. 217.

74 The figures come from an unpublished questionnaire by Van de Loo among 103 clinical psychologists, 61 of whom had filled in the form and returned it (Archives of the History of Dutch Psychology, Groningen).

75 J. H. Dijkhuis, *Klinische Benadering en Klinische Psychologie* (inaugural lecture, State University of Utrecht) (Utrecht: Erven J. Bijleveld, 1965), p. 9.

76 J. H. Dijkhuis, Het proces van de interactie tussen psycholoog en cliënt (public lecture, State University of Utrecht) (Utrecht: Erven J. Bijleveld, 1963), p. 7–8.

77 Verslag resultaten enquete, *De Psycholoog* 16 (1971):437–59.

78 G. Krijnen, *Ontwikkeling Functievervulling van Psychologen: Een Onderzoek naar de Ontwikkeling van Functies die door Psychologen in de Periode van 1954 tot 1972 zijn Uitgeoefend* (Nijmegen: Instituut voor Toegepaste Sociologie, 1975).

79 The German quote (from Bertold Brecht) can be translated as "A good person, who would not want to be one?" J. T. Barendregt, "Klinische Psychologie is meer dan Therapie Bedrijven" (1974), in *De Zielenmarkt*, pp. 24–39 (p. 33).

80 For example, see J. T. Barendregt, "Klinische Psychologie is meer dan Therapie Bedrijven"; J. T. Barendregt, "De Relatie van Diagnostiek en Therapie binnen het Fobieënproject," *De Psycholoog* 9 (1974):295–308.

81 J. T. Barendregt, "Klinische Psychologie is meer dan Therapie Bedrijven" (p. 132).

82 J. T. Barendregt, ibid., p. 129. A little further on he says, however: "As Eysenck did little justice to Freud, so the above does not do full justice to Eysenck. With his unrelenting advocacy of scientific values he has contributed as few others to the emancipation of clinical psychology."

83 J. T. Barendregt (undated), "Bindmiddelen voor de Ziel," first published in *De Zielenmarkt,* pp. 77–95 (p. 77).

84 Following the title of a 1976 article included in J. T. Barendregt, *De Zielenmarkt,* pp. 46–58, under the title "Zielsvervuiling" (pollution of the soul).

85 For the history of the phobias project, see K. A. Soudijn, et al., "De Geschiedenis van het Fobieënprojekt," in I. A. M. H. van Krogten, et al., eds., *Fobieën: Het Amsterdams Fobieënproject* (Deventer: Van Loghum Slaterus, 1984), pp. 12–27.

86 J. T. Barendregt, *Karakters Van en Naar Theophrastus* (Deventer: Van Loghum Slaterus, 1977), p. 41. The "meaningless syllables" refer to a popular operationalization among memory researchers. They have test subjects learn meaningless syllables by heart in order to exclude variation caused by different knowledge beforehand. Such research is often accused of being irrelevant to memory functioning in life outside the experimental situation. The psychologist/methodologist Karel Soudijn expressed the opinion that Barendregt had in fact described *himself* thirty times. Barendregt agreed with this, but said that it applied especially to the portraits of the farmer, the blunderer, the yesman, the complainer, the suspicious, the miser, and "perhaps what you would never guess," especially the meek soul (source: letter from Barendregt to Soudijn dated 25 December 1977).

87 J. T. Barendregt, "Moge de Crisis Blijven" (1977), in *De Zielenmarkt,* pp. 145–9 (p. 147–8).

88 J. T. Barendregt, "Moge de Crisis Blijven" (1977), in *De Zielenmarkt,* pp. 145–9 (p. 149).

89 For details see T. A. Veldkamp and P. van Drunen, *Psychologie als Professie,* Chapter 10.

90 Sociologists have described this development as part of what they call the "collectivization" of care. In this process informal help by neighbors and family gradually becomes the task of specific professional groups. To this end professional associations are set up; they can first be joined by election and at a later stage by following training courses prescribed or organized by the association. The professional groups simultaneously aim for recognition, title protection, and financial support from the state. In this way in the course of time they monopolize the field that they have made their own. See C. Brinkgreve, et al., *Sociologie van de Psychotherapie I: De Opkomst van het Psychotherapeutisch Bedrijf* (Utrecht: Het Spectrum, 1979); A. de Swaan, *In Care of the State: Health Care, Education, and Welfare in Europe and the USA in the Modern Era* (Cambridge: Polity Press, 1988).

91 H. J. M. Hermans, *Waardegebieden en hun Ontwikkeling: Theorie en Methode van Zelf-confrontatie* (Amsterdam: Swets & Zeitlinger, 1974); H. J. M. Hermans, "Interactieparadigma versus experimenteel paradigma: de cliënt en de psycholoog als evaluerende instanties van een therapie," *De Psycholoog* 10 (1975):4–11.

92 A. P. Cassee, *Van Psychodiagnostiek naar Gedragsanalyse* (public lecture, Free University of Amsterdam, 1973).

93 A. de Swaan, *In Care of the State*, pp. 210–16.

94 See also C. Brinkgreve, *Psychoanalyse in Nederland*, p. 217.

95 A. de Swaan, "Hetze en Hybris: De Val van het Psychotherapeutisch Bedrijf," *Maandblad voor Geestelijke Volksgezondheid* 35 (1980):734–43.

96 For an explanation of how these figures were obtained and for more figures, see C. Dijkstra and J. Bannenberg, "Er zijn teveel Klinisch Psychologen," *De Psycholoog* 21 (1986):178–82.

97 For example, F. Verhage, "De Competentie van de Psycholoog in de Medische Situatie," *De Psycholoog* 13 (1978):523–34 (pp. 533–4); R. J. Takens, "De Klinisch Psycholoog," in R. A. Roe, ed., *Wat Doet de Psycholoog?* (Assen: Van Gorcum, 1984):144–80.

98 R. J. Takens, "Klinische Psychologie als Professie," in A. P. Cassee, M. Höweler, and R. H. C. Janssen, eds., *Inleiding in de Klinische Psychologie* (Deventer: Van Loghum Slaterus, 1981):347–73 (p. 370).

99 W. K. B. Hofstee, "De Methodische Competentie van de Psycholoog," in P. J. D. Drenth et al., eds., *Psychologie in Nederland: Enkele Ontwikkelingen in 1982* (Lisse: Swets & Zeitlinger 1983):25–41.

100 S. Schagen, "Het Evalueren van Psychotherapie: Een Alledaagse Bezigheid," *De Psycholoog* 16 (1981):22–36.

101 Including: S. Schagen, "Meten Met Matige Maten: Over Uitkomstcriteria bij het Onderzoek naar Psychotherapie," *Nederlands Tijdschrift voor de Psychologie* 34 (1979):65–85; S. Schagen, "Het Evalueren van Psychotherapie"; F. Beenen, "Enkele Knelpunten in de Psychotherapie-research," *De Psycholoog* 13 (1978):62–67; F. Beenen, "Indicatiestelling voor Psychotherapie: Deskundige Logica of Deskundologica?" *Nederlands Tijdschrift voor de Psychologie* 34 (1979):23–38; K. A. Soudijn, "Heeft de Psychotherapeut een Achilleshiel?" *De Psycholoog* 14 (1979):117–27; K. A. Soudijn, "Diagnostiek en Psychotherapie," *De Psycholoog* 15 (1980):450–9; S. Schagen, *Het Effect van Psychotherapie: Meetbaarheid en Resultaten* (Deventer: Van Loghum Slaterus, 1983); K. A. Soudijn, *Kwaliteit van Psychotherapie* (Meppel: Boom, 1982).

102 L. K. A. Eisenga et al., eds., *Psychologie en Psychotherapie* (Amsterdam: Nederlands Instituut van Psychologen, 1981).

103 R. W. T. Diekstra, S. Schagen, K. A. Soudijn, and H. R. Wijngaarden, "Psychotherapie en het NIP," *De Psycholoog* 18 (1983):389–401. See also K. A. Soudijn, *Kwaliteit van Psychotherapie*.

104 K. A. Soudijn, "Diagnostiek en Psychotherapie."

105 For example, J. H. M. Ettema, "De Waardering van Kwalitatief-diagnostische Methoden," in B. G. Deelman et al., eds., *Ontwikkelingen in de Klinische Psychologie* (Deventer: Van Loghum Slaterus, 1986), pp. 37–61.

106 See B. G. Deelman, et al., eds., *Ontwikkelingen in de Klinische Psychologie;*

J. A. M. Winnubst et al., eds., *De Metamorfose van de Klinische Psychologie: Nieuwe Ontwikkelingen in de Klinische en Gezondheidspsychologie* (Assen: Van Gorcum, 1991).

107 For instance, see P. M. G. Emmelkamp, *Phobic and Obsessive Compulsive Disorders: Theory, Research, and Practice* (New York and London: Plenum Press, 1982); P. M. G. Emmelkamp, "Behavior Therapy with Adults," in S. Garfield and A. Bergin, eds., *Handbook of Psychotherapy and Behavior Change* (New York: Wiley, 1986); M. A. van den Hout, "The Explanation of Experimental Panic," in S. Rachman and J. D. Maser, eds., *Panic, Psychological Perspectives* (Hillsdale: Lawrence Erlbaum, 1988), pp. 237–58.

108 J. T. Barendregt, "Fobieën en Verwante Angsten," in *De Zielenmarkt*, pp. 163–81.

109 J. T. Barendregt and N. H. Frijda, "Cognitive Aspects of Anxiety," *Tijdschrift voor Geneesmiddelenonderzoek* 9 (1982):17–24.

110 J. T. Barendregt, "Fobieën en Verwante Angsten."

111 N. H. Frijda, "In Memoriam Prof. Barendregt," *De Spiegoloog* 1 (1982), p. 20.

Chapter 5. Predictions

1 For Laudan's discussion with sociologists of science, see J. R. Brown, ed., *Scientific Rationality: The Sociological Turn* (Dordrecht: Reidel, 1984); L. Laudan, *Sciences and Values: The Aims of Science and Their Role in Scientific Debate* (Berkeley: University of California Press, 1984).

2 L. Laudan, *Sciences and Values*, pp. 33–4.

3 The first eleven editions, as well as the English edition, were published by Mouton in The Hague, whereas the twelfth edition was issued in 1994 by Van Gorcum in Assen. The English translation of De Groot's book was not very successful. A review of this translation appeared in *Contemporary Psychology* (16 [1971]:538) that said "its approach is a European pedagogic 'Here's how science is done.' " De Groot built up a strong international reputation in another way. His book *Thought and Choice in Chess* (The Hague: Mouton, 1965) still belongs, according to the Science Citation Index, to one of the 0.1% most quoted studies (See: K. J. Vincente and W. F. Brewer, "Reconstructive Remembering of the Scientific Literature," *Cognition* 46 [1993]:101–28).

4 See the English-language edition: A. D. de Groot, *Methodology: Foundations of Inference and Research in the Behavioral Sciences* (The Hague: Mouton, 1969), pp. 27–33.

5 In the English translation of *Methodologie* this aphorism is given as: "If one knows something to be true, he is in a position to predict; where prediction is impossible, there is no knowledge" (*Methodology*, p. 20). As noted in Chapter 1, because the Dutch social scientists learned to say it in the first-person singular, I use a more literal translation here.

6 W. K. B. Hofstee, *De Empirische Discussie: Theorie van het Sociaalwetenschappelijk Onderzoek* (Meppel: Boom, 1980), p. 17

7 P. Kampschuur and K. Soudijn, "Interview met A. D. de Groot," *Psychologie* 3 (1984):12–18.

8 G. Visser, *Profiel van de Psychologie* (Muiderberg: Coutinho, 1985), p. 60.

9 A. D. de Groot, *Methodology*, p. 78.

10 Ibid., p. 19.

11 Ibid., p. 90.

12 Ibid., pp. 327–57.

13 W. K. B. Hofstee, *De Empirische Discussie*, p. 19.

14 Ibid., p. 24.

15 Ibid.

16 For a more extensive English-language explanation of the betting model by Hofstee himself, see W. K. B. Hofstee, "Methodological Decision Rules as Research Policies: A Betting Reconstruction of Empirical Research," *Acta Psychologica* 56 (1984):93–109.

17 Under the entries *voorspellen* or *voorspelling* (i.e., predicting or prediction) in the index of De Groot's *Methodologie* and Hofstee's *Empirische Discussie*, numerous references and even just *passim* are found; whereas in most English-language logical empirical standard works, "prediction" produces considerably fewer references. See, for instance, C. G. Hempel, *Aspects of Scientific Explanation and Other Essays in the Philosophy of Science* (New York: The Free Press, 1965); E. Nagel, *The Structure of Science* (London: Routledge & Kegan Paul, 1979); H. Reichenbach, *Experience and Prediction* (Chicago: The University of Chicago Press, 1937).

18 C. G. Hempel, *Philosophy of Natural Science* (Englewood Cliffs, NJ: Prentice Hall), p. 37.

19 A. D. de Groot, *Methodology*, p. 19.

20 C. G. Hempel, *Philosophy of Natural Science*, p. 19.

21 See, for example, C. Sanders, L. K. A. Eisenga, and J. F. H. van Rappard, *Inleiding in de Grondslagen van de Psychologie* (Deventer: Van Loghum Slaterus, 1976), pp. 88–9; C. Sanders and J. F. H. van Rappard, *Tussen Ontwerp en Werkelijkheid* (Meppel: Boom, 1982), pp. 96–7; W. K. B. Hofstee, "Boekbespreking van G. H. de Vries 'De Ontwikkeling van Wetenschap,'" *Nederlands Tijdschrift voor de Psychologie* 40 (1985):301–4.

22 K. R. Popper, *The Logic of Scientific Discovery* (New York: Harper & Row, 1959; repr. 1967).

23 An intriguing example of this is the broadening of the term "new" or "novel fact" in critical rationalist circles. Critical rationalists, including Popper, have taken the position that the qualification "novel fact" can also be lent to phenomena that researchers already described before, but for which a new explanatory theory has not been specially designed. A fact can be called novel and be a valid corroboration of this new theory as long as it has played no role in the *creation* of that theory: Then the theory "predicts" the fact. Naturally, in such a case there is no question of a prediction in De Groot's sense of foretelling future still-unknown research outcomes (see H. Zandvoort, "Criteria voor succesvolle wetenschappelijke researchprogramma's," *Kennis & Methode* 7 (1983):47–71 (p. 59–60).

24 K. R. Popper, *The Logic of Scientific Discovery* (repr. 1967), p. 105.

25 A. D. de Groot, *Methodology*, p. 66.

26 Ibid., p. 78.

27 Ibid., p. 81.

28 In De Groot's view too, human decisions are taken in science. For this he introduced the idea of "the tribunal of the history of the particular scientific discipline" that is "the forum of expert opinion" (*Methodology*, pp. 22–7 and pp. 107–13). This is because the problem of induction, combined with the fact that in the social sciences probabilist hypotheses are generally formulated, implies that most hypotheses and theories can never be rejected or established with certainty and that therefore they must be decided on in a different manner (*Methodology*, section 4). At first sight, this seems to be an important similarity to Popper and for this reason, too, it has been claimed in the Netherlands that De Groot is a critical rationalist. However, the *level* on which decisions are discussed differs. Whereas De Groot, in relation to *empirical* statements emphatically allots no function to the decisions of "the forum," in this context Popper always speaks of "decisions" and emphasizes that there can be no certainty. In Popper's own view there is "a vast difference" between himself and those who differ with him on this point especially: "For I hold that what characterizes the empirical method is just this: that the convention or decision does not immediately determine our acceptance of *universal* statements but that, on the contrary, it enters into our acceptance of the *singular* statements that is, the basic statements" (*The Logic*, repr. 1967, p. 109). Although in choosing a theory, according to Popper, no decisions have to be taken, because of the conventional character of the basic statements in principle *nothing definite* can ever be said about theories. For De Groot the uncertainty applies only to *most* hypotheses or theories. If neither the problem of induction nor probabilism is present, that is, in dealing with hypotheses on a domain that as a whole can be researched, or with deterministic hypotheses, these can be *definitively* established or rejected respectively (*Methodology*, section 4).

29 These books are L. Festinger and D. Katz, *Research Methods in the Behavioral Sciences* (New York: Holt, Rinehart, and Winston, 1953); C. W. Brown and E. E. Ghiselli, *Scientific Method in Psychology* (New York: McGraw Hill, 1955); A. Kaplan, *The Conduct of Inquiry: Methodology for Behavioral Sciences* (1964, reprint New York: Intertext books, 1973); B. B. Wolman and E. Nagel, eds., *Scientific Psychology: Principles and Approaches* (New York: Basic Books, 1965); H. Zetterberg, *On Theory and Verification in Sociology* (Totowa: The Bedminster Press, 1965); J. Galtung, *Theory and Methods of Social Research* (Oslo: Scandinavian University Books, 1967); H. M. Blalock and A. B. Blalock, *Methodology in Social Research* (New York: McGraw Hill, 1968); J. M. Neale and R. M. Liebert, *Science and Behavior: An Introduction to the Methods of Research* (Englewood Cliffs, NJ: Prentice Hall, 1973); F. N. Kerlinger, *Foundations of Behavioral Research* (New York: Holt, Rinehart, and Winston, 1973); F. N. Kerlinger, *Behavioral Research: A Conceptual Approach* (New York, Holt, Rinehart, and Winston, 1979); K. D. Opp, *Methodologie der Sozialwissenschaften* (Reinbeck bei Hamburg: Rowohlt, 1976); S. Nowak, *Methodology of Sociological Research* (Dordrecht: Reidel, 1977); M. Lewin, *Understanding Psychological Research* (New York: John Wiley, 1979). M. Mitchell and J. Jolley, *Research Design Explained* (New York: Holt, Rinehart, and Winston, 1988).

30 And the same holds for so-called controlled experiments and even for double-blind designs. Taking care that neither the experimenter nor the subjects know who gets which stimulus does not yet imply that researchers predict research outcomes.

31 If the Dutch prediction-rule were the *same* as testing null hypotheses, De Groot and Hofstee could have restricted themselves to an exposition of significance testing. However, they did not even choose it as a point of departure for the rest of their argument. De Groot discusses the prediction rule for all types of research and discusses the null hypothesis only in passing (in some cases the confirmation value of a test outcome can be determined with this, he writes on page 201 of *Methodology*). Hofstee even presents his betting model as an alternative to testing null hypotheses. Testing null hypotheses, he argues, sets a premium on taking little risk and only yields statements on what is *not* the case (W. K. B. Hofstee, *De Empirische Discussie*, pp. 52–9).

32 Kaplan also stated that in "most cases errors of observation can neither be prevented nor canceled out. What is possible is to *discount* the error, make ourselves aware of its direction, and perhaps even of its extent, and take it into account in our treatment of the observational data" (A. Kaplan, *The Conduct of Inquiry: Methodology for Behavioral Sciences*, pp. 129–30).

33 C. W. Brown and E. E. Ghiselli, *Scientific Method in Psychology*, p. 40.

34 K. D. Opp, *Methodologie der Sozialwissenschaften*, pp. 377 and 389–90.

35 G. Heymans, *Psychologie der Vrouwen* (Amsterdam: Wereldbibliotheek, repr. 1910), p. 30.

36 G. Heymans, "Presidential Address" (1926), in *Gesammelte kleinere Schrifte zur Philosophie und Psychologie* (Den Haag: Martinus Nijhoff, 1927, serial nr II), p. 366.

37 J. L. Prak, "Het Bedrijfsleven en de Toegepaste Psychologie," *De Opbouw, Democratisch Tijdschrift* 10 (1927–8):726–45 (p. 737).

38 J. L. Prak, Ibid., pp. 737–8.

39 H. J. F. W. Brugmans, *Over de Waardebepaling der Psychotechniek* (Groningen: J. B. Wolters, undated), pp. 14–15.

40 H. J. F. W. Brugmans, *Psychologische Voorlichting bij de Beroepskeuze* (Groningen: J. B. Wolters, 1920), p. 23.

41 J. Luning Prak, *De Moderne Onderneming en haar Personeel* (Amsterdam: Kosmos, 1947), p. 79. As mentioned earlier, in this period Prak changed his name to Luning Prak.

42 J. Luning Prak, ibid., p. 76.

43 A. D. de Groot (personal communication). In that time no general distinction was made, as is now the case, between the logical empiricist and critical-rationalist philosophy of science. Against his will, Poppers's *Logik der Forschung* (1934) was read as a specific elaboration of the logical empiricist program. It was a full twenty years later, after the appearance of the English translation of *The Logic of Scientific Discovery* (1959), that people began to characterize, for example in introductions to philosophy of science, "empiricism" and Poppers's "rationalism" as opposing movements (for this see G. H. de Vries, *De Ontwikkeling van Wetenschap* [Groningen: Wolters Noordhoff, 1984], p. 49).

44 A. D. de Groot, *Het Denken van den Schaker* (thesis, University of Amsterdam) (Amsterdam: Noord-Hollandse Uitg.Mij., 1946).

45 In the foreword to the first edition De Groot himself wrote: "Although in 1956 the emphasis was still on psychology, and psychodiagnostics in particular . . . in the course of the years the notes for and the successive textual versions of the book in hand acquired an increasingly general-methodological character. . . . Thus, while this book has been written by a psychologist, it has become a *methodology for the behavioral sciences.*"

46 A. D. de Groot, *Methodology*, p. 19.

47 Chapter 3 deals with the context of these ideas.

48 D. J. van Lennep, *Gewogen, Bekeken, Ontmoet in het Psychologisch Onderzoek* (inaugural lecture at Utrecht) (The Hague: Martinus Nijhoff, 1949).

49 D. J. van Lennep, ibid., p. 28.

50 B. J. Kouwer, *Gewetensproblemen van de Toegepaste Psychologie* (inaugural lecture at Groningen) (Groningen: Wolters Noordhoff, 1955), included in B. J. Kouwer, *Persoon en Existentie* (Groningen: Wolters Noordhoff, 1977), p. 59.

51 B. J. Kouwer, *Gewetensproblemen*, p. 55

52 W. K. B. Hofstee, *Psychologische Uitspraken over Personen* (Deventer: Van Loghum Slaterus, 1975).

53 W. K. B. Hofstee, *De Empirische Discussie*, p. 22.

Epilogue. Social and rational rules

1 Hornstein argues something similar. Her answer to the question of how American psychological research became quantitatively oriented is that control, prediction, and classification served the needs of a growing bureaucracy and the democratic ideology of American institutions (G. Hornstein, "Quantifying Psychological Phenomena: Debates, Dilemmas, and Implications," in J. Morawski, ed. (*The Rise of Experimentation in American Psychology* [New Haven and London: Yale University Press, 1988], pp. 1–35). Geuter reaches comparable conclusions with regard to the methods in early German psychology (U. Geuter, *The Professionalisation of Psychology in Nazi Germany* [Cambridge: Cambridge University Press, 1992]). And Smith shows that physiology and physiological psychology are, through their language on inhibition, bound up with social evaluations of being human (R. Smith, *Inhibition: History and Meaning in the Sciences of Mind and Brain* [London: Free Associations Press, 1992]).

2 G. W. Stocking, "On the Limits of 'Presentism' and 'Historicism' in the Historiography of the Behavioral Sciences," *Journal of the History of the Behavioral Sciences* 1 (1965):211–18 (p. 217). Stocking uses the concept of historicism, but maybe *historism* should be preferred to avoid a connotation with naive empiricism and, moreover, to avoid confusion with the term historicism used in Karl Popper's book, *The Poverty of Historicism* (London: Routledge and Kegan Paul, 1957), where this term refers to the idea that the course of history is subjected to strict laws. This is an idea that historism (or historicism in Stocking's sense) actually opposes.

3 That pedantry can possibly be mitigated somewhat by emphatically stating that whatever is said in this book about the methodology of the social sciences can be applied equally well to my view of the methodology of science studies and the present project itself. That can be said, as I will make clearer toward the end of this epilogue, without undercutting the basis of this study. Therefore I will not go any further into the subject. Others, however, have devoted much attention to the "reflexivity" of science studies. See, for example, S. Woolgar, ed., *Knowledge and Reflexivity: New Frontiers in the Sociology of Knowledge* (London: Sage, 1988); M. Ashmore, *The Reflexive Thesis: Writing Sociology of Scientific Knowledge* (Chicago: Chicago University Press, 1989).

4 See also I. Hacking, "The Accumulation of Styles of Scientific Reasoning," in Dieter Henrich, ed., *Kant oder Hegel? Ueber Formen der Begruendung in der Philosophy* (Stuttgart: Klett Cotta, 1981), pp. 453–65. And I. Hacking, "Language, Truth, and Reason," in M. Hollis and S. Lukes, eds., *Rationality and Relativism* (Oxford: Basil Blackwell, 1982), pp. 48–66.

5 Laudan calls this the "hierarchical model of justification" and he rejects it on similar grounds. As an alternative – and in opposition to the "new wave" sociologists of science – Laudan subsequently develops a "reticulated model," which exchanges the idea of a universal rationality for that of a relative rationality. What is rational is not measured against general standards, but is constantly reasoned again by comparing what scientists say they want to achieve with what they actually accomplish and how they go about it (L. Laudan, *Science and Values: The Aims of Science and Their Role in Scientific Debate* [Berkeley: University of California Press, 1984]). In this model there is no room for an implicit or explicit discussion on social and political matters among scientists and nonscientists, carried on simultaneously with the methodological debate. In these cases under the regime of the reticulated model, therefore, the epithetet "scientific" can not be granted. However, as Laudan himself emphasized when he rejected the hierarchical model, methodology is underdetermined by cognitive aims and is therefore debatable. If this is so, then on what grounds can a restrictive measure as to the *participants and subjects* of the methodological debate be based? Rather than the sciences that express a social identity in their methodology, it seems to be the reticulated model that should be dismissed.

Index

Milton Keynes UK
Ingram Content Group UK Ltd.
UKHW041522181024
449640UK00009B/153

9 780521 144872